Digital Disruption and Corporate Digital Responsibility

Michaela Bednárová

Logos Verlag Berlin

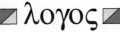

Reviewers:

Prof. Enrique Bonsón, PhD.
Doc. Ing. Zuzana Kubaščíková, PhD.

Bibliographic information published by the Deutsche Nationalbibliothek

The Deutsche Nationalbibliothek lists this publication in the Deutsche
Nationalbibliografie; detailed bibliographic data are available
on the Internet at http://dnb.d-nb.de .

ISBN 978-3-8325-5496-5

Logos Verlag Berlin GmbH
Georg-Knorr-Str. 4, Geb. 10,
D-12681 Berlin
Germany

Tel.:   +49 (0)30 / 42 85 10 90
Fax: +49 (0)30 / 42 85 10 92
https://www.logos-verlag.com

Table of Contents

Introduction

The publication "Digital Disruption and Corporate Digital Responsibility" is an original monothematic work of scientific-exploratory character where I critically analyse, evaluate, and synthesize the knowledge related to new disruptive technologies, Artificial Intelligence (AI) ethics and disclosure, and the possibility of corporate continuous reporting. In addition, the results of my own scientific work related to Automated Decision-Making (ADM) disclosure practices; AI disclosure standards; as well as a continuous reporting model are presented. It is addressed to the scientific community dealing with new technologies in accounting and reporting.

In the last decade, technological disruption in businesses has been driven by two main forces, Artificial Intelligence (AI) and blockchain (BC). This work focuses on analysis and implications of both efficient data collection and processing via AI and its underpinning technologies, and its secure storage via blockchain.

According to the Gartner survey (2019), 37% of organisations have implemented AI to some extent and the number of businesses using AI globally, has tripled from 2018 to 2019 (Gartner, 2019). The main reason for this uptake is that most of the AI technologies and other developments that create potential for AI systems, such as robotics or Internet of Things (IoT), have matured enough to be ready for wide-scale adoption (Benaich and Hogarth, 2020).

Over the last few years, EU investments into AI have increased substantially and the EU objective is to increase these investments, up to 20 billion euros per year over the next ten years (Benaich and Hogarth, 2020).

Regarding blockchain, 347 million euros of EU funding have already been designated to blockchain research and innovation projects in the following areas: security; public services; Internet of Things; sustainability (production, traceability, circular textiles, energy and transport); advanced manufacturing; 5G; AI and Big Data; food security; innovation support; media and social media (European Commission, 2022).

Nevertheless, Europe's ambition to be a global leader in adopting the latest technologies is not limited to profitability and efficiency. It is also important to highlight the benefits and promote the development of secure, sustainable, inclusive and trustworthy AI, which would respect human rights (European Commission, 2021b).

As significant investments in AI in businesses are expected over the next decade, there is a need for transparency, ethics, and responsibility in this area as the costs of non-action could be high. Therefore, in this work, one of the questions I aim to answer is how to guarantee the trustworthiness of AI in business processes.

Whether, and to what extent, a company integrates new technologies in its business processes is difficult to assess without access to internal corporate information. Eventually, companies might disclose this information voluntarily on their website, annual report or even in the press, as it might appeal to investors and other stakeholders. Nevertheless, whether to trust that the company acts in compliance with AI ethical principles is left to subjective judgement. Therefore, developing AI ethical standards is just the first step. Although a proposal for AI regulation is the next important step forward, it still does not cover the need for transparency in this matter, which might have a significant effect on stakeholder decision-making. Thus, to fill this gap, this work also aims to shed more light on current corporate AI disclosure practices and contribute to standardization in this emerging area.

Last but not least, the world has changed and reporting must too. Therefore, in addition, an idea of a reporting model that would meet the needs of the 21st century and would take advantage of the latest technological development is outlined.

The present work consists of three main parts: Digital Disruption; Corporate Digital Responsibility; and Continuous Reporting. Additionally, a brief section is dedicated to discussion about the changing profile of the accounting and auditing profession.

In the first chapter, an overview of AI and its related technologies such as Big Data, IoT, and Machine Learning (ML) is provided with a closer look at their definition, main functionalities, implications for accounting, and challenges. In terms of blockchain, a comprehensive technology description, brief overview of current projects, as well as the implications and limitations for accounting and auditing are discussed.

The second chapter deals with the ethical implications of AI usage. Therefore, the concept of Corporate Digital Responsibility is discussed and results of my research on AI disclosure and standardisation are presented.

The third chapter discusses a powerful combination of AI, blockchain and other relevant technologies such as XBRL and a framework for continuous reporting is presented.

The fourth chapter outlines the changing profile of the accounting and auditing professions.

The methods applied in this work are a comprehensive literature and technology overview; automated content analysis of corporate reports; focus groups; and questionnaires regarding the most important aspects for AI disclosure.

Thus the scope of this publication can be summarised as follows:

1st part: Digital Disruption:

- discusses current trends in new disruptive technologies (development, increase of usage by businesses, investments),
- provides technology description, definitions, implications, and limitations or challenges of AI; Big Data; IoT; ML; and blockchain.

2nd part: Corporate Digital Responsibility:

- provides an overview of AI ethical standards and initiatives,
- provides an overview of EU initiatives and regulation,

- introduces and discusses an emerging concept of Corporate Digital Responsibility (CDR),
- provides an overview of previous research,
- introduces my contribution to the field (related to ADM disclosure and AI disclosure framework),
- explains the methods applied in my previous research in this area,
- discusses the need for an AI disclosure framework,
- outlines future research and limitations.

3rd part: Continuous Reporting:

- discusses the importance and possibility of continuous reporting,
- outlines principles of continuous reporting,
- introduces the semi-automated continuous reporting model.

4th part: Changing Profile of the Account and Auditing Profession:

- discusses the diversification of the accountant role,
- discusses the emergence of new sectors due to technological disruption,
- outlines changes in the accounting and auditing curriculum.

1 Digital Disruption

The recent technological disruption or fourth industrial revolution (Schwab, 2016) is driven by two main forces, Artificial Intelligence and blockchain.

1.1 Background

AI refers to a program or a set of programs or algorithms that are able to automate processes but also reproduce some features of human behaviour. AI incorporation into business processes has led to cost savings, increased turnover, and higher quality of products and services.

Blockchain is a distributed digital registry, where information is recorded and shared through a peer-to-peer network. Thus, one of its most important features is decentralisation, which means that ledgers are stored on different nodes instead of a single location, and each authorised participant has an identical copy. Therefore, any changes to the ledger are reflected in all copies almost immediately, which makes the recorded data immutable and the blockchain a distributed and secure database.

As significant investments are expected to flow to both AI and blockchain projects in the following years, in this chapter we are going to look closer at the functionality of the technology, implications for accounting, and challenges.

1.2 Big Data, Internet of Things, Artificial Intelligence and Machine Learning

1.2.1 Big Data

Definition, benefits, and usage

As Franks (2012) reported, according to the McKinsey Global Institute (2011), Big Data are datasets, the size of which is beyond the ability of typical database software tools to collect, store, manage and analyse. Other intrinsic features of Big Data are as follows: data is automatically obtained/generated; is not formatted for easy usage; and is of little use unless it is structured (Franks, 2012; Moffitt and Vasarhelyi, 2013). Increasing processing power and storage capacities currently enable Big Data analytics to be used by medium and small companies as well. Companies seeking a competitive advantage have increasingly extended the scope of their information systems, replacing traditional data processing by automated data capture and procession. Previous research shows that Big Data can indeed increase profitability and that businesses that had embraced Big Data for decision-making process had 5-6% higher profitability (McAfee and Brynjolfsson, 2012). According to Moffitt and Vasarhelyi (2013) Big Data can be used in different sectors, i.e. finance and insurance, to analyse risk and detect fraud; the utility sector, to analyse usage and detect anomalies; or marketing, to analyse customer behaviour.

Data processing

Big Data can exist in different formats: structured; semi-structured (e.g., XML based); unstructured (e.g., text, phone calls, videos); and multi-structured (different types of data through different structural levels). Nevertheless, unstructured data are the most common and therefore represent the biggest challenge for its further processing and analysis (Moffitt and Vasarhelyi, 2013). Over the years, different techniques to standardize unstructured data have been developed, e.g. mathematical and machine learning techniques (Aizawa, 2003); latent semantic analysis

(Landauer et al., 1998), and cluster analysis (Thiprungsri and Vasarhelyi, 2011).

Implications for accounting

Accounting serves, on the one hand, to gather and report information useful for management, on the other hand, it discloses relevant information to a variety of external stakeholders. From the internal point of view, applying Big Data and Business Analytics can enforce internal controls, increase production efficiency, reduce waste or discover anomalities (Moffitt and Vasarhelyi, 2013). In addition, it has great benefits for marketing by collecting a wide range of data from customers such as click-path, GPS location, preferences, which enables more in-depth analysis of consumer behavioural patterns and allows microtargeting. From an external point of view, it can offer up-to-date information of different kinds to external stakeholders such as investors to help them make informed decisions without a time-lag. Thus, the Big Data approach can improve both internal and external accounting and reporting. Auditors can also benefit from Big Data usage as it would enable them to verify that a transaction took place with multi-modal evidence (including photos, videos, GPS location) (Vasarhelyi et al., 2010).

Weaknesses and challenges

Perhaps one of the main weaknesses of Big Data is the lack and difficulty of standardization.

Another important challenge relates to data protection. In particular, the EU is becoming stricter in this area by recently introducing the GDPR, which brought various projects related to Big Data to an end as their realization might have caused certain infringement with the new regulation.

Another threat is security and the possibility of data leaks. Last but not least, the processing power and storage capacities might also be challenging the wide adoption of Big Data in different sectors.

1.2.2   Internet of Things (IoT)

Definition, benefits, and usage

There are predictions that everything will be connected to the Internet in less than five years via more than 21.000.000.000 devices, of which 80% would be connected via non-conventional networks. Alternative networks such as Sigfox developed by Cellnet, which allows one way transfer of small portions of data for a low cost (e.g. a GPS location can be sent via Sigfox, but streaming a video via YouTube is not possible), might appear.

The application of IoT can cause changes in all areas and industries, e.g. home automation (a smart fridge that orders food), biotechnology (digital twins that allow predictions of body reactions on treatment), microtargeting (new marketing models on the individual level), automobile industry (different sensors, autonomous cars), intelligent agriculture (continuous monitoring of cultivation). The IoT is able to collect a huge amount of data of all kinds via different sensors for further analysis, therefore, the IoT enables Big Data to live.

Data processing

The key components of IoT are processors, sensors, and communication networks. Regarding processors, continuous evolution in this area would lead to the production of smaller processors that will be smaller and with low consumption. In terms of sensors, there are plenty of opportunities to design and develop sensors of all kinds.

Communication networks such as Sigfox are useful to boost IoT usage. Sigfox is an alternative to WiFi, but its main characteristics are: low consumption, serves to send a small portion of information such as GPS location, and enables one-way communication. Its coverage is mostly within Europe and USA.

Implications for accounting

Nowadays, we are witnessing a shift in the nature of accounting records and the incorporation of non-traditional sources of data into accounting, which challenges current accounting standards and opens new ways for accounting data collection and processing. Traditionally, accounting as the domain of business measurement has been restricted to financial reporting, where data collection and registration depended on the technological and cost/benefit considerations. With the change in technology, economic transactions can be traced and measured faster and more in depth. In addition, the possibility to measure non-financial indicators via different sensors allows companies to identify, measure and act on sustainability issues more easily. The increased requirements placed on large corporations in terms of their sustainability performance is necessary, but causes an immense burden for the companies which have to identify, measure and report on those data. Therefore, the IoT and automated data collection would increase the efficiency of such data collection and reporting. Thus, incorporation of Big Data via IoT into the accounting process would increase the volume, velocity, and variety of accounting data (Vasarhelyi et al., 2015) with important implications for management, financial accounting, and sustainability reporting. The business environment is full of real-time data and via IoT and the development of new technology we have the opportunity to obtain them. Thus, the current accounting system based on quarterly or annually reporting, measurements such as FIFO/LIFO, measurements based on the historical cost, or estimates for annual depreciation are neither up-to-date nor accurate. Therefore, Vasarhelyi et al. (2015) proposed the alternative accounting measurements based on technology development (Figure 1).

Figure 1 Traditional versus technology-enhanced accounting measurement

Traditional versus Technology Enhanced Accounting Measurement

Traditional Measurement	Updated Measurement	Technology Used	Accounting Concept Needed	Big Data Extensions
LIFO or FIFO inventory valuation	Current value of held inventory	Values of items in electronic markets, sales lists, infomediary price lists	Inventory revaluation adjustment to equity	RFID
Historical cost of PP&E	Current value[a] of PP&E	Search at different levels of aggregation (e.g. factory, machine part) market values	Partial valuation, geographic valuation differences, etc. Separation of different types of increase in organization net value	Peer benchmarks, extractions of histories of pricing and activity in peer markets
Depreciation	Current value comparisons across time	Values of items in electronic markets, sales lists, infomediary price lists	Partial valuation, geographic valuation differences, appreciation, etc.	
Intangibles	Reports on human resources, intellectual property, supply chain, customer reactions, service support obligations	Utilization of the ERP applications as HR, IP, SC	Additional information that may or may not link to financial statements; may or may not be dollar denominated	Social media, news pieces, labor market extracts with salaries, etc.

[a] Although there is a rich literature of price-level accounting and different bases of measurement (SFAS No. 33, replacement cost, current value, etc.), technology at that stage was not cost efficient to enable these measurements to be practical.

Source: Vasarhelyi et al. (2015)

Challenges

The use of IoT is in constant conflict with data protection rights. In particular, a strict GDPR law of the European Commission has stopped many projects related to the IoT due to data privacy intrusiveness. Another challenge is security and the threat of data leaks. Distributed denials of service attack (DDoS) which compromise the system by sending an immense amount of orders and manage to block the service worry both the companies providing the service and the authorities.

Other challenges are related to the lack of standardization in this area. Thus, questions might arise such as: (1) what type of non-traditional data should be collected (video, audio, text, GPS coordinates, etc.); (2) what would be the optimal frequency of data collection; (3) how much data should be retained; (4) what is the appropriate level of data aggregation.

In addition, in the near future, we might expect many applications to provide some level of bridging between the data source and financial applications (Vasarhelyi et al., 2015).

1.2.3 *Artificial Intelligence (AI) and Machine Learning (ML)*

Stein (2018) studied the implications and applications of more continuous accounting and reporting. He points out the role of artificial intelligence as one of the driving changes and potential disruptors in the accounting profession. Numerous projects of Big 4, but not limited to them, related to the accounting implications of AI have already started to dramatically shift the landscape of both accounting and auditing.

Definition, benefits, usage

AI refers to a set of programs or individual program that is able to reproduce some features of human behaviour. As it becomes more sophisticated over time, the capacities and consequences of this technology continue to increase (Stein, 2018). The main ramifications of AI in the accounting would be more automation in accounting work, in particular, in lower level tasks where early career professionals currently spend an immense amount of working hours and more efficient use of continuous data. This disruption would also lead to job adjustments or even job eliminations. Hence, a debate about how to update the skillset of future accountants and auditors is in place.

Being able to design tools for the efficient collection and correct interpretation of business data is becoming a strong competitive advantage. Increased digitization of information requires substantial investments, but many businesses are aware of the choice to "innovate or perish".

Within the wide range of artificial intelligence applications, it is *Machine Learning* that might provide the greatest impact in business. Machine learning is a technique that provides systems the ability to automatically learn and improve from experience. It has a potential to enable fast and intelligent decisions with an accuracy and efficiency beyond the human capabilities. It focuses on the development of computer programs that can access, analyze data and learn from them.

Data processing

Machine learning is usually fed by Big Data in which it identifies certain patterns and inference and is able to make predictions by using algorithms and statistical models. The most common categories of machine learning algorithms are supervised, unsupervised, semi-supervised and reinforcement ML algorithms. The choice of a particular algorithm depends on the nature of the data. If it is an existing training dataset we use a supervised ML algorithm. In contrast, when the information is neither classified nor labelled, unsupervised ML algorithms are applied, which are able to explore the data and draw inferences from datasets. Reinforcement ML algorithms interacts with the environment and are based on a trial and error approach to determine the ideal behaviour within a specific context[1].

---

1   https://expertsystem.com/machine-learning-definition/

Implications for accounting

As ML is able to perform complicated tasks with the level of accuracy and efficiency beyond the capacities of human employees, it can be used to maximize business productivity. For example, in the financial sector, it would provide the ability to profile customers more accurately; in the production process, it might help to identify faults and errors; in the HR, it could make recruitment or staff retention automated, more accurate and efficient.

Although internal reporting has become very rich in terms of large sets of information of different kinds such as production, marketing, supply chain, human resources, etc., companies rarely take full advantage of it. Those data are often stored separately, while it might be interesting to measure mutual relationships among them. Here, AI could play an important role as it would allow a substantive increase in data analysis and predicting capacities.

Challenges

The wide array of challenges related to AI usage include the economic, technological, regulatory, management and ethical aspects. One of the main challenges of economic character is to obtain initial investments and have sufficient resources and tools to implement and maintain AI elements. Another challenge is related to algorithm accuracy, which is on the one hand a technological challenge and on the other hand, it may have some ethical implications. Therefore, the algorithms designed should guarantee non-bias and non-discrimination as delegating decisions to intelligent machines may have profound moral and ethical implications. Regarding management, there is an upskilling need in information communication technology (ICT) skills, where both management and employees operating the services should have a necessary set of skills. In recent reports, experts point out the need for a complex curriculum which would cover ICT, business and transversal skills to satisfy the current labour market needs (CHAISE, 2021b). Last but not least, regulatory oversight might impede development and use of some AI, in particular those which are invasive, such as facial recognition.

1.3 Blockchain

Definition, benefits, and usage

In recent years we are seeing important technological advances, one of which is blockchain, which is considered by many to be the most disruptive technology since the internet (Swan, 2015). Currently, numerous projects are being launched with the aim of exploring the potential of this technology in different sectors, far surpassing the interest it initially aroused in the financial and technological sector (Bonsón and Bednárová, 2018).

Although blockchain was created as a support technology for a cryptocurrency (Nakamoto, 2008), it has given rise to a wide range of applications that take advantage of its characteristics to record and manage a variety of information.

Blockchain is a distributed digital registry system that allows information to be recorded and shared through a peer-to-peer network. Identical copies of the ledger are maintained and validated by members of the network. The information enters into blocks that are added to a chain chronologically and are connected through a cryptographic validation called "hash".

As we can see in Figure 2, each new block has its timestamp and contains information that refers to the previous block. This chaining between blocks is done through the hash and each block has the hash reference of the previous block. Therefore, any alteration to a block would change the hash. Thus, any attempt to modify the records in the blockchain would require the alteration of each previously created block, which is unlikely given the decentralized nature of the technology (Bonsón and Bednárová, 2018).

Figure 2   Structure of blocks

Source: Tenomad (2018)

According to Palfreyman (2015), blockchain increases transparency and auditability, since all transactions are visible to each participant in the architecture. Transactions are stored in multiple places, which can improve access to information (Swan, 2015). Due to the pre-established consensus to add a transaction in a block (Mainelli and Smith, 2015), control is also increased. This leads to higher data trust and reliability as transactions are verified by multiple nodes.

According to Palfreyman (2015) and Tapscott and Tapscott (2016) blockchain can reduce the costs of executing and validating a transaction through automatic verifications and leads to higher data integrity and quality. At the same time, Cai and Zhu (2016) point to the reduction of human errors due to automatic transactions and controls.

In addition, according to Tapscott and Taspcott (2016), this technology could solve a current problem with data privacy, where all the information we provide about ourselves with each online movement creates a virtual footprint which is not our property. It is observed, therefore, that blockchain could solve this problem by cryptographically organizing our identity in a "black box" that would allow us to share only the information necessary for each particular transaction. Palfreyman (2015) refers to this phenomenon as the resolution of the "privacy paradox" and states that blockchain can offer effective identity management solutions.

Data processing

As explained previously, one of the most important features of blockchain is decentralization, where ledgers are stored on different nodes instead of a single location, and each authorized participant has an identical copy of the ledger. Any changes to it are reflected in all copies almost immediately. The data once recorded is immutable, which makes the blockchain a distributed and secure database.

Thus, blockchain as a decentralized registry allows: better control of information and transactions; the ability to carry out a transaction without the intermediation of a third party; greater resistance to malicious attacks and accidental failures; greater transparency and quality of information (Bonsón and Bednárová, 2018).

The design of the blockchain network architecture rests on four key decisions related to control, data ownership, privacy, and access (Ølnes et al., 2017). We can therefore differentiate between a public and private network, taking into account who can access copies of the ledger, and crossing that choice with the type of centralized or decentralized network, depending on who maintains the network (Mainelli and Smith, 2015).

In the private-permissioned blockchain, a private blockchain is licensed, so participants need permission to join. Therefore, only trusted organizations can transact and participate in the consensus process. Another category is the private-permissionless type, whose main characteristics are restrictions on access and transactions, but there are no limitations on participation in the consensus mechanism. The public-permissioned design is probably best suited for government organizations as there are no restrictions on data access or transactions. However, there are restrictions on participation in the consensus process. In the public-permissionless architecture there are no limitations on access, transactions, or validation. Therefore, it is the architecture that most closely resembles the initial vision of Blockchain technology (Bonsón and Bednárová, 2018).

The private blockchain network seems to be the solution that most enterprise applications would lean towards. Its design is based on a closed group of designated trusted validators that verify and execute transactions. Participants need an invitation or permission to join, and the control mechanism may vary. In some cases, the existing participants are the ones who can decide the future participants. In others, it is the regulatory authority that issues participation licences, and a third option is when the decision is made by a consortium. Once an entity has joined the network, it will play a role in maintaining the blockchain in a decentralized manner (Jayachandran, 2017).

Some authors, such as Gandía (2018), add the hybrid or consortium blockchain to the classification, differentiating it from public and private ones. In hybrid blockchain, the blockchain is semi-public with limited access. The consortium is composed of a restricted number of participants who are authorized according to the control mechanisms of the network. However, the information in the general ledger is public, with which there is control over who can intervene while providing public transparency to all the information.

With blockchain comes the concept of Distributed Automated Organization (DAO), which represents an organization whose governance and operations are executed in blockchain environments capable of functioning without human involvement. As Buterin (2014) points out, the autonomous function of DAO companies is based on cryptotechnology and automation, under the control of a trustworthy set of programmable rules. Certainly, the concept of DAO may seem somewhat utopian until it is contrasted with enough real experiences. However, it shows that it is feasible to evolve towards this system in some organizations through blockchain or, at least, the automation of some processes can be incorporated to improve efficiency. Smart contracts and autonomous agents could be considered the first steps towards that goal.

Smart contracts are intelligent contracts that are self-executing in the case of meeting pre-established conditions. Buterin (2014) defines them as systems that automatically move digital assets according to previously specified rules. Interestingly, it was Szabo in 1994 who began to introduce smart contracts, but its application did not prosper until the development of distributed database technology. This is so because without blockchain a third party was needed to supervise and execute the contract, while with this technology the automated execution of smart contracts became feasible as the supervision responsibilities were distributed among the participating nodes (Dai and Vasarhelyi, 2017). With blockchain, smart contracts can be used more easily compared to the technology available at the time of its invention (Nofer et al., 2017).

Smart contracts are contracts capable of executing themselves autonomously and automatically. They are based on predefined business logic that is agreed upon by the contractual partners and once configured, it can be programmed and stored in the blockchain ledger (Rozario and Vasarhelyi, 2018). Consequently, users activate the smart contract by sending data to it, and the smart contract verifies the received inputs against predefined rules and releases an output. Figure 3 shows an example where a blockchain-based smart contract is used to monitor and operate a loan agreement.

Figure 3   Example of a smart contract

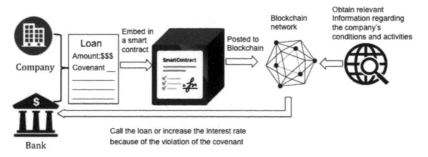

Source: Dai y Vasarhelyi (2017)

Implications for accounting

Blockchain seems to be an interesting tool that can cause improvements in different industries, as it offers solutions for greater data auditability, control, and reliability. The application of this technology would reduce not only costs, but also human errors by automating transactions through smart contracts and, more importantly, it would help prevent manipulation and fraud. Due to its numerous advantages, this technology has not gone unnoticed in the audit industry. Consequently, we find that the Big Four have initiated different projects with the aim of exploring their potential in the fields of accounting and auditing (Minichiello, 2015; Allison, 2015).

One of the most concise definitions of blockchain is to consider it a distributed ledger which highlights the proximity of this technology to accounting. In 2005, thus, before the invention of blockchain, Ian Grigg titled his design based on distributed ledger technology (DLT) as Triple Entry Accounting. Subsequently, the accounting literature highlighted the possibilities of blockchain and its application to accounting (Dai and Vasarhelyi, 2017).

Grigg (2005) starts from the premise that the single-entry or single-item accounting system has the drawback of struggling to detect errors and, therefore, there is a high risk of fraud. The double-entry method meant a revolution because the maintenance of the basic accounting equation makes the transactions be matched and leaves a trace on all the accounting facts. But the double entry system still needs proof or verification, which implies costs and duplication. Hence, the third entry provides advantages by containing the cryptographically signed receipt of the transaction, offering control by all parties, with the same information for users and issuers.

The verification, which for Grigg (2005) would have to be carried out by a neutral and trusted third party, for Dai and Vasarhelyi (2017) could be carried out exclusively by the blockchain through the consensual registry in the distributed ledger. In this way, a third layer starring the blockchain would be embedded in the current double entry system, which entails the shared registration of transactions. Figure 4 shows a purchase-sale transaction in the accounting information system, where the blockchain is incorporated into the traditional double-entry model.

Every transaction is recorded on the distributed ledger independently of traditional double-entry accounting records. The system generates a stream of data on the blockchain that is stored through token transfers. In turn, it makes it possible to give different views of the information depending on the type of users and the tokens in the distributed ledger would serve as certificates of ownership of assets or obligations contracted between the collaborating companies. The example in the figure refers to the purchase of goods on credit: the transaction is recorded in accounting and, simultaneously, a token is transferred to the blockchain ledger for storage and monitoring purposes, with the identification information, related transactions, current balance, and cryptographic keys for verification.

The application of smart contracts in the system would allow instant compliance verification with the basic accounting equation, but also maintain inventory permanently, order payments or any other function that is programmable and executable such as detecting errors or duplicities, check that the records comply with accounting standards or agreements between companies (for example, the discount for prompt payment could be established based on the time it is made, at the time the events occur) (Dai and Vasarhelyi, 2017; Wanden-Berge et al., 2019).

In blockchain any asset can be tokenized and with it:

- is assigned a digital identity,
- can be transmitted over the network,
- traceability can be followed,
- its components and characteristics can be registered.

The tokens implemented in the blockchain can simply be means of payment (cryptocurrencies) or represent goods or rights for the consumption of goods or services (utility tokens), as well as shares or securities (security tokens).

Therefore, a transaction between two companies implies, as is logical, the corresponding double-entry accounting entries in accordance with the accounting normalization of the moment and simultaneously, in a connected way, a record in a third book in the blockchain. This record in the blockchain would be cryptographically sealed, agreed and registered in all

Figure 4   Representation of a sale-purchase transaction in blockchain

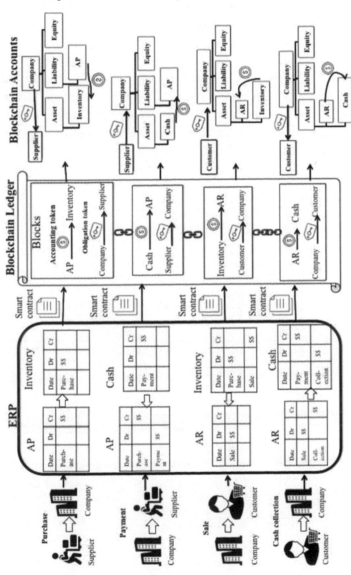

Source: Dai y Vasarhelyi (2017)

the nodes of the network, with its consequent time stamp. Hence, the condition of unalterable is attributed to this registry, with the certainty that the information has not been modified nor can it be altered. In parallel, as the assets are tokenized, their traceability could be established and actions automated with the execution of smart contracts, such as ordering payments when certain conditions are met, giving alerts on inventory status, issuing invoices, warnings about treasury balances, confirmation of accounts receivable or payable balances. and any other process that can be scheduled. Another aspect to consider is the possibility of making the information contained in the blockchain visible to different users, and this is closely related to the design of the network. In short, it seems that a new ecosystem for accounting information is evolving (Dai and Vasarhelyi, 2017; Kokina et al., 2017).

Table 1 depicts how the quality of accounting information can be improved by blockchain.

Table 1   Information quality dimensions in blockchain landscape

Information Quality Dimensions (IFRS)	Information Quality Improvement by BC
Completeness (IFRS)	Information has to be complete (the requirements of completeness are predefined by approved nodes).
Interpretability and clarity (IFRS)	Due to the predefined requirements for completeness, the interpretability and clarity of information is increased. Each entry of the blockchain has predefined fields which have to be filled in. This facilitates the interpretation of information.

Relevancy (IFRS)	Different access levels are possible. Some nodes, such as CEO of the company or auditing company might have access to all information while other stakeholders might have limited access (only aggregated information will be displayed) based on their predetermined roles. Some content might be available to users who have the decryption key. Hence, each node has access to the information which is relevant to it.
Comparability (IFRS)	Due to the certain standardization of predefined fields, information of similar nature might be easily compared.
Other information quality dimensions	Information Quality Improvement by BC
Authority	The source of the information is recognized. Only preselected (trusted) nodes of the blockchain ecosystem can insert the information. Additionally, the "author", is easily traced and identified.
Accuracy	Information is verified by nodes or smart contracts. Accuracy is guaranteed as various nodes have to agree in order to add it.
Timeliness	Blockchain allows for instant information updates, therefore, continuous "on-time" reporting is possible.
Manipulation	Almost impossible. Read and write permissions are restricted to certain entities. Once the information is added into a block it is cryptographically locked and immutable.

Source: Bonsón and Bednárová (2019)

Challenges

Although many authors point out the variety of benefits that this technology can bring, many of them are not supported by sufficient empirical evidence since blockchain is still in its infancy.

On the other hand, Dai and Vasarhelyi (2017) present three different contexts of challenges that can hinder the adoption of this technology. In the technological context, they refer to the financial and time resources required by the technical complexity of blockchain, therefore, a company could face difficulties finding commercial partners (and other entities) with whom it could share a decentralized architecture. The organizational context is marked by the willingness of managers to accept the establishment of this technology. Therefore, the perceived benefits must outweigh the potential costs, while the environmental context points to the essential role of regulators in the adoption of blockchain within the accounting ecosystem.

From a technical point of view, programmers and designers have to face many challenges to bring the projects to maturity. It is worth noting that in order to design suitable blockchain solutions for accounting purposes, several aspects must be taken into account, such as the selection of nodes, the structure of the database, the identifications, the verification protocols, etc. (Bonsón and Bednárová, 2018).

Thus, to fully integrate it into a real business ecosystem, both technical and non-technical aspects such as consensus between regulators, auditors and public administration must be resolved (Bonsón and Bednárová, 2018; Dai and Vasarhelyi, 2017).

Our research related to blockchain disruption of industries

Given its disruptive potential, after conducting the literature review and desk research related to blockchain implications for accounting and auditing (Bonsón and Bednárová, 2019), our further research focuses on an analysis of the level of blockchain disruption in different sectors (Bonsón et al., 2022b). For that, a content analysis of corporate annual reports (or sustainability reports) has been conducted. The sample consisted of listed European companies operating in 11 sectors. Large companies are those who have sufficient financial resources to engage in innovative projects. An automated content analysis using R-programming was conducted to identify companies which include any mention of the blockchain initiative in their corporate reports. Thus, the reports have been "scanned" by looking for keywords such as "blockchain", "distributed ledger technology (DLT)" or "smart contracts". The preliminary results show that over the last few years, engagement with blockchain was limited to the financial and technological sector. Even now, most of the companies involved in blockchain projects come from this sector. Nevertheless, other sectors seem to be catching up. For example, one of the most common applications of blockchain outside of the financial and technological sector seem to be innovative traceability solutions in supply chain and sustainability (Bonsón et al., 2022b).

1.4   Powerful Combination of IoT, AI and Blockchain

While each of these technologies presents exciting opportunities and significant challenges to any organization, the combination of all three can be truly transformational.

To understand the powerful combination of the Internet of Things (IoT), artificial intelligence (AI), and blockchain, it is helpful to imagine it as a set of interconnected organic processes. IoT is like the human nervous system, sensing situations through connected devices. AI is like the reasoning part of the brain, which thinks and makes decisions by analyzing data, and blockchain is like memory, which creates a secure and immutable record of transactions and data exchange.

IoT is transforming a world of things into a world of data. Virtually anything can be equipped with a sensor, from a smartwatch that monitors heart rate and blood sugar levels, to a connected factory that monitors every stage of the production process. There is a huge opportunity for businesses in using IoT, but the challenge is to leverage existing technology, infrastructure, and capabilities to accelerate performance while minimizing the cost and complexity of deployment (Wanden-Berghe et al., 2019, 2020).

Within the broad field of artificial intelligence, machine learning is poised to have the biggest impact, since it can enable fast and intelligent decision-making, either supporting human intelligence or replacing it. Through machine learning, companies can delegate complicated tasks to achieve a level of accuracy and efficiency beyond the human capabilities.

Blockchain can represent an efficient solution anywhere digital information is exchanged. In the consumer goods sector, blockchain will provide transparency throughout the supply chain through asset tracking, improving accountability, streamlining product, and improving consumer trust. Whether in education and research or in the music industry, it will help ensure that intellectual property rights are respected. And of course, in the realm of finance, blockchain will empower financial technology (Wanden-Berghe et al., 2019).

IoT, machine learning, and blockchain are potentially transformative on their own, but exponentially more powerful when combined. Figure 5 shows the evolution of IoT networks from a closed system to a completely decentralized and open system that will use "clouds" for data storage (Hedgething, 2016).

Figure 5 Blockchain and IoT

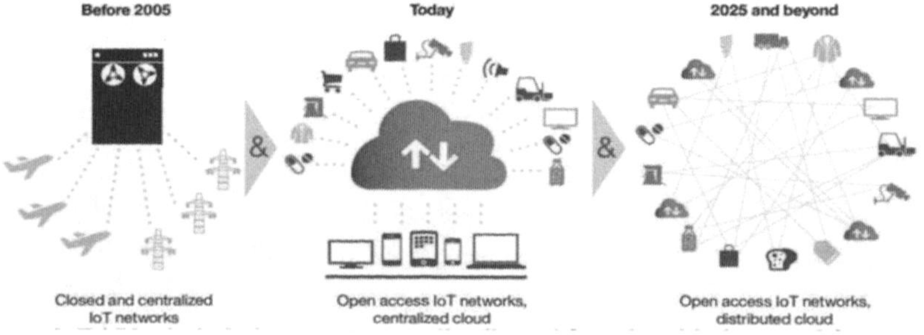

Source: Hedgethink (2016)

Figure 6 IoT, blockchain, and AI

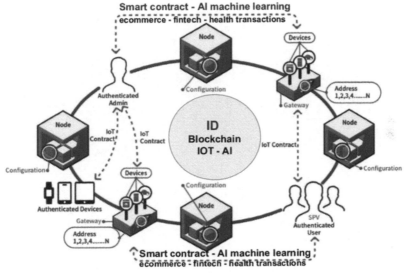

Source: Guarda (2018)

Figure 6 presents the combination of IoT, blockchain, and artificial intelligence: data is collected through IoT, stored and managed through blockchain, and analysed and interpreted through artificial intelligence. In this way, in artificial intelligence, the smart contract could be combined with machine learning, allowing future situations and behaviour to be predicted.

As Dai and Vasarhelyi (2017) illustrates, the manufacturing industry can use a cloud solution to monitor production where those responsible for production could remotely control each machine in the factory. This way, they can have real data about the performance of each equipment and through machine learning, which would be able to anticipate a failure of the machine, even proactive maintenance is possible. Thus, by incorporating these technologies, the system would not only be able to track and monitor the activities of physical objects in real time, but also automate some business processes.

2 Corporate Digital Responsibility

In this chapter a new concept, Corporate Digital Responsibility, will be introduced and discussed.

2.1 Background

Over the last decade we have been witnessing a rapid evolution of new technologies whose incorporation into business processes has led to significant cost savings, increased turnover, higher quality of products and services, and higher job satisfaction. This trend is expected to continue and increased efficiency due to AI adoption will become even more evident in the upcoming years.

According to Gartner (2019), AI is a key investment for many businesses globally as the adoption of AI has tripled from 2018 to 2019. Investments are mainly funnelled into robotic process automation, business process management software, or machine learning.

However, large-scale incorporation of AI into business processes might raise certain concerns related to privacy, data protection, or other human rights which might be threatened when AI is used for decision-making, in particular, when a high level of automation in decision-making might contain inherit biases.

Therefore, rapid proliferation of these new technologies are reasons for the general public to worry, leading to increased stakeholder scrutiny and recent regulatory interest in this matter.

Despite the recent EC's initiatives to increase the trustworthiness of AI such as Ethical Guidelines for Trusted AI, published in 2019, or the White Paper on AI: A European Approach to Excellence and Trust (2020) that have laid the foundations for the European regulation of AI, Artificial Intelligence Act, published in April 2021, recently, there is no legislation requiring companies to be transparent about their AI responsibility. Yet, some companies have responded to direct or indirect stakeholder pressure by voluntarily disclosing such information, although, the reason for such disclosure seems to appeal more to company investors.

As AI usage might represent significant potential risks, there is an urgent need for their mitigation via requiring corporate transparency as well as verification of disclosed information in this area.

When it comes to company transparency beyond financial disclosure, perhaps we can see a parallel between AI disclosure and non-financial disclosure related to environmental and social issues which have been in the spotlight over the last twenty years. We have witnessed how significantly companies evolved in the way of presenting information about environmental, social and governance aspects. It started with stakeholder pressure due to the possible negative impacts of company activities on the environment and society. Thus, companies have been gradually asked to be more transparent and to complement their financial statements with non-financial disclosure, which led to the creation of generally accepted non-financial reporting standards such as the Global Reporting Initiative (GRI), EMAS, ISO 26000, SA 80000, etc.

What started on a voluntary basis soon transformed into legislation. The effort started in 2014 when the European Council, as one of the earliest promotors of non-financial reporting globally, on October 22, adopted a new Directive 2014/95/EU on the disclosure of CSR by large companies and groups, which modifies the previously adopted Directive 2013/34/EU on annual financial statements, consolidated financial statements, and related reports of certain types of companies, followed by the publication of the Commission Communication 2017/C215/01 related to guidelines for the presentation of non-financial reports.

Thus, AI disclosure might follow a similar path, starting with stakeholder questions about data privacy, protection and human rights, followed by disclosure standards and regulation.

Recently, the European Financial Reporting Advisory Group (EFRAG) was appointed to elaborate EU non-financial reporting standards (EF-

RAG, 2021). In line with this call, the project Task Force within the European Corporate Reporting Lab adopted a multi-stakeholder approach with the main aim of proposing a roadmap for the development of a comprehensive set of EU sustainability reporting standards. In February 2021, the final report was published containing 54 proposals. It is important to mention that the final report focuses more on the scope and structure of possible EU sustainability standards than on specific disclosure requirements, indicators or metrics. It builds upon a well-established ESG classification and recommends three layers of reporting: sector-agnostic, sector-specific and entity-specific.

For the purpose of our research, we explored more in depth the social sub-topics related to Human Rights and Governance sub-topics related to business ethics as these are the topics most closely related to our area of interest.

Proposal *#37* states that disclosure should be aligned with international and EU reference frameworks and standards, including the UNGP on Business and Human Rights, the OECD Guidelines (and other international declarations and principles), as well as the Charter of Fundamental Rights of the EU (Bonsón and Bednárová, 2022). The recommendation is rather general and a clear articulation of what should be reported in this category is missing. Nevertheless, as one of the risks related to AI usage is that it might challenge or interfere with certain human rights, AI disclosure could be considered more seriously within this particular proposal.

Proposal #38 deals with governance sub-topics, in particular, business ethics. Again, there are no specific instructions regarding data privacy for example, but this issue could be a part of this particular category.

Due to the strong relationship between AI usage and human rights or business ethics, the area of disclosure in this field is worth exploring as a possible future component of sustainability disclosure.

Current AI disclosure, being voluntary and considered more a novelty than a standard, is logically characterized by a lack of standardization.

Contemplating the importance of corporate transparency to mitigate risks related to AI usage and satisfy the information needs of different stakeholders, a significant effort should be made to reach a general consensus on what and how to report.

Therefore, apart from analysing the current corporate AI disclosure practices, one of our current research interests was a study related to proposing and validating a set of general elements that should be reported on to structure information about AI, which would improve transparency, mitigate risks, and prove a company's digital responsibility.

Thus, this research line can be considered one of the first attempts to map, analyse, structure and standardize corporate information related to AI where the analysis of current AI disclosure practices, together with a proposal of the AI reporting framework can contribute to both literature and practice. In terms of the literature contribution, this research can extend the scarce academic literature on AI disclosure. Regarding the practical implications, the results of this research line can help companies to get acquainted with current trends in AI disclosure as well as assisting them to structure AI information focusing on materiality principle (Bonsón and Bednárová, 2022).

2.2   Technology and Ethical Implications of its Usage

2.2.1   Artificial Intelligence vs. Automation

Before a discussion about which elements should be included in AI disclosure, a brief technology overview is provided.

AI is a subset of intelligent automation, and it helps automation to evolve. Thus, intelligence automation is a wider term, and covers a vast technology to conduct work tasks such as: Robotic desktop automation (RDA); Robotic process automation (RPA); Visual recognition technology; Artificial Intelligence (AI). AI includes Machine Learning (ML), predictive and adaptive models, natural language processing (NLP) and Deep Learning (DL) as a subset of ML.

Automation focuses on repetitive tasks and helps save time and cost on monotonous and voluminous tasks. AI, on the other hand can mimic human intelligence, decisions, and actions. In other words, AI technologies are able to imitate how human beings speak, hear, understand, react, and make decisions. Machine Learning (ML) as a part of AI, is able to learn and identify patterns in big data sets and recommend decisions without human involvement. Within ML we can find a powerful tool, Deep Learning (DL), as its subset. DL is based on artificial neural networks which learn continuously. In addition, it is able to adjust itself automatically and improve the accuracy of outputs.

There is a wide range of different AI tools. Supposedly, any AI that learns from big datasets on its own and is autonomous in making decisions is potentially dangerous. Nevertheless, when trying to impose transparency on AI, it is important to consider the scope of its usage and probability of implying any harm or ethical issues. Therefore, the area of application of highly autonomous AI is crucial to decide whether more transparency would be required as the biased algorithmic logic in sensitive areas might imply more harm to individuals than AI in i.e. manufacturing.

The example of potentially less harmful environment of AI application would be applications such as *textio*, which is an AI tool using predictive technology to write job listings to make them sound lucrative to potential candidates. Companies like Microsoft, Starbucks, or Twitter have been using it. *Conversica*, which is used to manage gross sales needs is another example of a "lower risk" AI app. Similarly, *X.ai* is a smart assistant for handling company meetings. Another AI tool in this category is *Digital Genius*, which is based on NLP and is able to maintain human like conversations with customers. *Intraspexion* regularly scans internal company emails and flags potential litigation risks for future review. Thus, it can prevent and warn companies against litigation risks. Similarly, a facial recognition app, which might be considered sensitive, but is used for photo tagging on social networks, can also belong to the "lower risk" category due to its area of application.

Then, there is another group of AI tools, which despite being based on the same technology (ML, NLP, etc.), is considered riskier due to its area of application. Examples might be AI tools used in human resources to select candidates or facial recognition technology used in litigation processes (AlgorithmWatch, 2019).

Hence, AI tools used in the areas with low potential harm might not automatically be considered risky. On the other hand, usage of autonomous AI tools, or "black box" algorithms in the areas with potential harm should be treated differently. For this reason, it is important that organisations not only disclose which technology they use but also where, as it is an important factor to assess a level of risk. (Bonsón and Bednárová, 2022).

2.2.2 Automated Decision-Making (ADM) Definition and Ethical Concerns

Automated Decision-Making Systems (ADM Systems) have their origins in the late 1950s, early 1960s (Power et al., 2019). One of the first attempts to conceptualize ADMs comes with Cyert and March (1963), who developed computer models for business decision-making in a company, referring to it as a decision system. A decision system can be defined as a set of interacting methods, people, procedures, programs, and routines to support decision processes or make the final decision (Power et al., 2019).

Power et al. (2019) stated that the research on decision systems as organizational phenomena has been diverse and different synonyms to define it have been used over the years. In addition, their study provided a comprehensive summary of research and definitions related to decision systems over the last six decades.

Simon (1960), Gorry and Scott-Morton (1971) identified two subcategories of computerized decision systems 1) automated decision systems, and 2) decision support systems. Gerity (1970) introduced a term "man-machine decision system" (MMDS), which referred to the interaction of three main components: man, machine and a decision task. Recently, terms such as automated decision system (ADS), decision automation system (DAS), or automated decision making (ADM) systems have appeared in the literature.

According to Costa (2021), ADMs are systems that sense data, apply algorithmic logic, and make decisions - all with no or minimal human intervention. He also states that ADMs can help companies reduce labour costs, improve quality, enforce policies, or even leverage scarce expertise, and respond to customers. Sangdon (2017) defines ADM as the ability to make decisions based on generated profiles without human intervention. He also pointed out that these processes entail the use of personal data and stressed the importance of personal data protection.

A number of previous studies show that the ADMs can be applied in different sectors including the public sector (Cobbe and Singh, 2020); manufacturing (Grishina, 2012); education (Mougiakou et al., 2019); healthcare (Davenport and Harris, 2005); telecommunications (Seufert et al., 2016); retail; financial services (Power et al., 2019), etc. The area of application might vary from automated pricing decision, loan approvals (Power et al., 2019), to automated driving (Gerwien et al., 2021).

As automating decisions proliferates into different sectors and industries, organizations should consider different organisational, technological, and legal aspects when deciding which decisions could be computerized to improve efficiency.

Cobbe and Singh (2020) consider reviewability of machine learning systems an important aspect to guarantee the accountability of ADM. They point out the importance of breaking down the ADM process into its technical and organisational components to gain a complete understanding of it.

Hence, to be able to fully understand and review the algorithmic logic of a particular ADM, company transparency is necessary. Therefore, our study aimed to shed some light on the extent to which companies disclose the use of using ADM it (Bonsón et al., 2022).

ADM systems are already used in many aspects of our lives including but not limited to credit scoring; college admissions or recruitment; welfare benefit eligibility; criminal justice decisions, or border control (Hickock, 2020).

Algorithmic decision systems (ADSs) or algorithmic decision making (ADM) are based on the analysis of big datasets to infer correlations or to derive information to make decisions. The same way as humans are prone to errors and might be biased, systems can also be biased. ADM is fed with data that might be biased. Therefore, if there is a bias in the data that the ADM is trained with, that effect can be reinforced and amplified, which might ultimately lead to unfair or even discriminatory decisions.

Thus, the decision yielded by ADM can be discriminatory without a company intending to discriminate. For example, gender or race are variables that normally cannot be taken into consideration when making a decision. Nevertheless, these variables might often be statistically associated with seemingly inoffensive characteristics, such as height or postal code. This is how an ADM which works with huge sets of correlated data can lead to indirect discrimination. Therefore, individuals should have some basic rights, such as transparency or the right to object, if there is a suspicion of biased decision-making by a machine/algorithm (Bonsón et al., 2022).

Therefore, it is necessary to know who built them, why (the main purpose), how they were developed, and how and where they are ultimately used. To improve algorithm accuracy and minimize bias, the algorithm should be fed with a representative dataset, the right model should be selected, and the algorithm should be reviewed periodically. The transparency of a company in these aspects might not only increase the general trust in ADM, but also serve as proof of a company's commitment to corporate digital responsibility (CDR).

2.3 International Initiatives Related to AI Principles and Ethical Standards

The hype around AI systems is accompanied by challenges related to ethical issues. This is drawing the attention of both academics and practitioners due to several reported cases demonstrating biased algorithms and their impact on individuals and society (Hickock, 2020).

Over the last few years we encountered several ethical incidents related to the AI-enabled systems. Perhaps the most well-known scandals were: (1) Amazon recruitment AI biased against hiring women (by mistake, the AI detected that mostly men were hired and learned this pattern), or (2) when Cambridge Analytica used data from Facebook users without their knowledge for political campaign purposes (Vakkuri et al., 2020).

As a response to current challenges related to AI ethical aspects, the publication of AI ethics principles and guidelines from civil society organizations, research centres, private companies, or governmental agencies emerged (Hickock, 2020). These institutions positioned themselves on what they consider are the most important values to be embedded in the development and implementation of AI products.

There are more than 100 AI ethics principles and guidelines around the world (Hickock, 2020). Algorithmwatch (2020) launched a database of AI Ethics Guidelines Global Inventory, which currently includes 167 guidelines. The most well-known are IEEE Global Initiative on Ethics of Autonomous and Intelligent Systems (2016), Future of Life Institute (2017), European Commission (2018). As a consequence of this trend, big players from the private sector have also developed certain ethical standards in this matter, e.g., Microsoft (2018), IBM (2018), Google (2018), or Telefonica (2018). Although the proposed principles might vary slightly, they are all based on common points such as: justice, transparency, non/maleficence, privacy, security, responsibility, respect for human rights (Bonsón and Ortega, 2019, Bonsón and Bednárová, 2021a, Hickock, 2020). Perhaps the most advanced initiative to regulate the area of AI is the one introduced by the European Commission, which is discussed in the next subchapter.

Despite the fact that these initiatives represent a step forward, the current system still has major flaws. The survey of Vakkuri et al. (2020) analysed 211 software companies in order to provide an overview of where we are in terms of AI ethics concluding that although some companies have developed initiatives related to good practices, given its voluntary character and giving it the emphasis this topic deserves, ethics in AI still has a long way to go (Bonsón and Bednárová, 2021b). In addition, Hickock (2020) points out the high level of abstractness of such principles and guidelines.

Bonsón et al. (2021a) analysed the top 200 Eurozone listed companies from Germany, Sweden, Finland, France, Spain, and Italy finding that despite the growing interest in AI, 43% of analysed companies were not reporting on any activity related to AI in their annual or sustainability reports. Less than 5% of them reported on some ethical approaches to AI and were mostly big companies from the technology and telecommunication sectors. Hence, the adoption of ethical approaches to AI are at a very initial stage.

2.4 EU Initiatives

On the following pages, EU initiatives such as the Coordinated Plan on AI (2018), Ethical Guidelines for Trusted AI (2019), and White Paper on AI: A European Approach to Excellence and Trust (2020) that have laid the foundations for the European regulation of AI, Artificial Intelligence Act, published in April 2021 are discussed.

2.4.1   The 2018 Coordinated Plan on AI

The EU aims to achieve global leadership on adopting the latest technologies. Thus, one of the latest commitments from the European Comission is to maximise Europe's potential in regards to Artificial Intelligence (AI) development and applications, encouraging member states to work together in this effort and help make the region a global reference. One of the first steps towards this commitment was the 2018 Coordinated Plan on AI, which defines a common direction, areas of investments, and strategic objectives for EU policy on AI (European Commission, 2018). Thus, the plan lays the groundwork for cooperation and points out the importance of developing national strategies on AI. At the same time, the EU promotes coordination instead of uncoordinated individual efforts of member states (European Commission, 2018, 2021b).

The key messages of the Coordinated Plan can be summarized as follows:

- Boost investments in AI technologies by accelerating private and public investments through EU funding schemas such as Digital Europe (DEP), Horizon Europe (HE) programmes and the Recovery and Resilience Facility (RRF).
- Implement AI strategies and programmes efficiently through joint efforts between Member States and different industries, taking advantage of digital innovation hubs (DIHs).
- Adjust AI policies to address global challenges (European Commission, 2021b).

2.4.2 Ethical Guidelines for Trusted AI (2019)

As a second step, after introducing the coordinated plan on AI in 2018, the European Commission set up an independent high-level expert group to formulate Ethics Guidelines for Trustworthy AI. Its main aim was to define guidelines to increase trustworthiness of AI systems. The guidelines revolve around three main components of AI throughout the system's entire life cycle: (1) it should comply with all applicable laws and regulations, (2) it should ensure adherence to ethical principles and European values, (3) it should be robust regarding its technical and social perspectives and should minimize unintentional harm (European Commission, 2019). Ethical guidelines point out four ethical principles: respect for human autonomy, prevention of harm, fairness, and explicability. Consequently, seven key requirements to be continuously evaluated throughout the AI system's life cycle were proposed: human agency and oversight; technical robustness and safety; privacy and data governance; transparency; diversity, non-discrimination and fairness; societal and environmental wellbeing; and accountability. Therefore, ethical guidelines represent a more detailed framework and served as a stepping stone for the White paper (European Commission, 2020).

In February 2020, the EU launched a white paper on "Artificial Intelligence – A European approach to excellence and trust" (European Commission, 2020) asking for feedback on their proposal on how to develop a regulative framework for AI systems. This initiative aims to address an increased demand from member states to create a common regulatory framework on which they can rely. Some separate initiatives have taken place in Germany, Denmark or Malta, therefore the EU assumed its responsibility and took action in this matter to avoid fragmentation in AI regulation and to work towards a solid common regulatory framework which would be in compliance with the general values of the EU (Bonsón and Bednárová, 2021b).

The main building blocks of this white paper are:

- Mobilising resources to achieve an **ecosystem of excellence** along the entire value chain where AI plays an important role. The ecosystem of excellence was introduced in the Coordinated Plan, which was part of the strategy on AI adopted in 2018. This plan aims to address societal and environmental well-being as a key principle for AI. Thus, the AI systems should help tackle current problems such as climate change.
- Building an **ecosystem of trust** by identifying the key elements of a regulatory framework for AI development and use. On one hand, AI can help protect citizens' security, on the other hand, citizens are concerned that AI can infringe their privacy or even be misused for malicious or criminal purposes. Thus, the factor of trust is crucial for a broader uptake of AI.

The feedback from different stakeholder groups is very important to identify which requirements related to the ethical and responsible use of AI are already reflected in existing legal or regulatory regimes and focus on those which are not. Generally, the requirements regarding transparency, traceability and human oversight are considered to be weak points as they are not specifically covered under current legislation. A clear European regulatory framework would help build trust in AI systems among citizens and companies and therefore might boost the uptake of this new technology (Bonsón and Bednárová, 2021b).

In addition, as the responsible development and use of AI can be a driving force to achieve the Sustainable Development Goals of the 2030 Agenda, the legal framework can be of great help as it will increase the confidence of both users and companies responsible for the development of AI systems. On one hand, this level of security might boost the uptake of AI tools, but on the other hand, it can also limit the research and development in this area as happened with a number of projects which had to stop when the GDPR was implemented. Thus, the EU must carefully balance the pros and cons while introducing the future legal framework so as not to create disproportionate obstacles and a burden for companies.

The first step would be to question whether the current legislation can address the risks of AI and whether those can be effectively enforced. The second step would be to analyse whether adjustments of the current legislation are needed, or whether new legislation should be developed.

The idea behind the regulatory framework is to focus on how to minimise the potential risks AI entails. The risks might have material (safety and health, damage to property) or immaterial character related to the privacy protection concerns, freedom of speech and expression, human dignity, or a different form of discrimination (Bonsón and Bednárová, 2021b).

Thus, the white paper draws attention to the two main categories of risks:

- **Risks for fundamental rights, including personal data and privacy protection and non- discrimination**

As AI enables the tracking and analysis of habits, there is a potential risk that state authorities, other entities, even employers, can observe how an individual behaves. Some of the technologies are able to de-anonymise data about individuals, hence, data privacy might be at stake or even used for malicious purposes. In addition, discrimination bias might have a much larger effect when controlled by a machine. This can happen when the AI learned while in operation and concluded a biased result. For example, certain algorithms designed for predicting criminal recidivism displayed gender and racial bias (Tolan et al., 2019).

The inherent features of AI technology such as complexity, unpredictability, and autonomous behaviour can make it difficult to verify compliance with existing EU law mechanism and may therefore inhibit an effective enforcement of rules.

- **Risks for safety and the effective functioning of the liability regime**

Uncertainty related to this type of risk can include who should be held responsible when an autonomous car which wrongly identifies an object on the road causes an accident. The legal uncertainty for businesses in this matter may reduce levels of safety but also discourage companies from AI development, which might undermine the competitiveness of European companies.

Scope of a future EU Regulatory Framework

First of all, the regulatory framework should clearly define what belongs to the scope of AI-enabled systems, therefore a definition would have to be provided. The Commission aims not to be excessively restrictive so it will not create a disproportionate burden, in particular for small companies. Thus, it proposed a risk-based approach where only high-risk AI applications should be regulated (Bonsón and Bednárová, 2021b).

According to the presented white paper, an AI application is considered high-risk if it meets two cumulative criteria:

- The AI application is used in a sector which implies higher risk such as healthcare, transport, energy and some areas of the public sector. The list of the sectors should be regularly updated.
- The AI application is used in such a manner that significant risks are likely to arise, for example here we would distinguish between the impact of a wrongly scheduled appointment in a hospital and an AI

application that might cause the risk of injury, death or significant material or immaterial damage.

Overview of general recommendations

In May 2020, AI Ethics community from *montrealethics* provided their feedback on the White Paper where they highlight, among others, the following points:

- GDPR includes the formula of the "right to object" in the data protection area. A similar approach could be adopted with AI where the "right to negotiate" related to the AI system's decision or output should be guaranteed.
- The use of facial recognition technology should be banned to avoid the risks related to the discriminatory issues and infringement of fundamental rights.
- All AI systems, regardless of whether they are low, medium, or high-risk, should follow similar standards and compulsory requirements.
- The biometric identification technology should be used only to fulfil the purpose for which it was developed and at the same time it should represent the best possible solution to achieve that purpose.
- A labelling system, a sort of certification, could be developed on a voluntary basis for the AI systems that are not considered high-risk.

In addition, the following aspects could be considered:

- If the AI system handles a huge number of data, it must be transparent about data governance.
- Companies should not gather any data which are not useful right now for a specific purpose with the intention that that data might be useful in the future.

2.4.3 *Proposal for Regulation (2021)*

In this subsection, the Commission's proposal of new European Union legislation to regulate artificial intelligence (AI) is briefly introduced and discussed. The proposal only marks the beginning of the legislative process and the European Parliament can still modify the proposal substantially (European Commission, 2021a).

In April 2021, the European Commission proposed a legal framework, which is supposed to serve as a proposal for regulation on a European approach for AI. The main goal is to guarantee the safety and fundamental rights of people and businesses within the scope of a Europe fit for the Digital Age. Hence, it defines rules and actions for excellence and trust in AI. The framework is based on a risk approach distinguishing four categories of AI systems: 1) unacceptable risk; 2) high-risk; 3) limited risk; and 4) minimal risk. Depending on the level of risk the AI system represents, the company would be obliged (or not) to provide a certain level of transparency. To start with, AI systems from an unacceptable risk category (applications and AI systems that could cause physical/psychological harm, manipulate human behaviour or exploit human vulnerabilities, or any application that would allow social scoring or real time remote biometric identification in public places) will automatically be banned. High-risk AI systems would require comprehensive risk management, which would guarantee quality of data, accuracy, robustness, traceability of results and cybersecurity. With limited risk AI systems, AI users should be at least aware that they are interacting with an algorithm. Thus, they will be subject to lower transparency requirements. For minimal-risk AI systems, e.g. AI systems in video games or spam filters, no disclosure will be required (Bonsón and Bednárová, 2022).

2.5 Research Related to ADM Disclosure and AI Disclosure Framework Development

2.5.1 Introducing Corporate Digital Responsibility (CDR) as a Part of CSR

As outlined previously, companies should approach the usage of high-risk AI responsibly as quality of data, accuracy, robustness, and traceability of results. Cybersecurity must be guaranteed. Hence, it would require comprehensive AI risk management. Lobschat et al. (2021) argue that algorithmic decision-making should follow the same ethical norms as human decision-making. We should not omit this ethical responsibility when incorporating new technologies. By assuming that technology usage implies certain responsibilities we get to a relatively new concept of corporate digital responsibility (CDR). This concept started to appear in the literature very recently and the scope of studies related to CDR is still narrow (Liyanaarachchi et al., 2020; Isensee et al., 2020; Lobschat et al., 2021). Liyanaarachchi et al. (2020) analysed CDR in the banking sector. Organizational structure, sustainability and digitalization in small and medium companies was analysed by Isensee et al. (2020). The study of Lobschat et al. (2021) was the first wider study on CDR where they introduced a framework of CDR culture. Despite an increasing interest in this concept among professionals, there is still a lack of conceptualization of CDR.

The Institute of Consumer Policy defined the following scope of CDR: data and algorithmic decision making; participation and reduction of inequality; digital education; future of work; digitalization in service of an ecologic transformation (Conpolicy, 2021). In the CDR conceptualization by Lobschat et al. (2021), companies should demonstrate their responsible behaviour in the digital age. They also drew some links between CSR and CDR.

CSR certainly shares some principles with CDR (Bonsón et al. 2022). Indeed, recent studies conducted by Deloitte (2019), Gärtner et al. (2018), PWC (2019) state that CDR is an additional layer of CSR and that early adopters of CDR disclosure include CDR-related information in their sustainability reports. Chief innovation officer at Deloitte, Nicolai Andersen, pointed out the importance of including CDR into the CSR scope (2019).

Over the last two decades, our attention has turned to CSR related to environmental, social and governance responsibility, but the recent digital revolution implies new responsibilities, which should be treated accordingly. Indeed, digitization affects almost all aspects of our lives (Gärtner et al., 2018) and therefore, companies taking advantage of this phenomenon should act responsibly.

On the other hand, different stakeholders also put pressure on companies to be more transparent about the activities that were previously undertaken by humans, but now different algorithms are involved in decision making. Therefore, it is natural that stakeholder pressure highlights the issues that could affect them and require more transparency. Some companies have started to react on that request and voluntarily disclose information related to new technologies, particularly if AI and ADM are involved. Nevertheless, such a disclosure is still in its very early stage, which implies a lack of uniformity. Hence, to take advantage of CDR, certain standardization in its reporting should be sought, similar to CSR. For that, a consensus should first be reached on which elements should be included and how to report on them.

2.5.2   *Previous Research on AI, AI Ethics and Disclosure*

Ethics in IT is not a new phenomenon, authors like Mason (1986) and Moor (1985) were the first academics discussing ethics in information technologies. Hence, the first studies related to information technology responsibility appeared in the early 1990s. In the following years, emerging IT was a source of continuous concern for more than the next two decades. Thus, ethics in its usage was the centre of attention of several academics (Bynum, 2001; Chatterjee et al., 2015; Floridi, 2010). Similarly, privacy concerns and data protection were the topics of several studies (Ashwort and Free, 2006; Beke et al., 2018; Culnan and Bies, 2003; Kehr et al., 2015). The unstoppable IT evolution and development of artificial intelligence and its use in business and public administration caused even more concern and AI ethics became the main topic of several authors (Angwin et al., 2016; Cath, 2018; Fjeld et al., 2020; Floridi, 2019; Greene et al., 2019; Hagendorff, 2020; Jobin et al., 2019; Robbins, 2020; Vidgen et al., 2020).

The comprehensive analysis of AI research in accounting by Sutton et al. (2016), where they compiled 30 years of observations, confirmed that AI proliferation in businesses continues to grow. Therefore, ethical implications continue to be an ongoing concern.

Although there are several studies dealing with the AI ethics, there are only few studies focusing on AI transparency via corporate disclosure. Bonsón et. al (2021) conducted one of the first studies focusing on AI activities and ethical approaches in leading listed companies in the European Union analysing corporate disclosure.

To fill this gap, in our recent line of research, we focus not only on analysing the ADM disclosure of large listed European companies, but we also aim to propose the first framework for AI-ADM disclosure (Bonsón and Bednárová, 2021b, 2022).

2.5.3   Theoretical Background

As AI disclosure in general is considered a part of CDR, which is a new layer of CSR disclosure, a question arises. Which theories could explain the motivation of companies to voluntarily disclose information related to AI usage? To fully understand this phenomenon, we have to understand a wide array of AI implications which are both economic and socio-political. Therefore, a multi-theoretical approach is proposed to be able to provide a comprehensive explanation for AI voluntary reporting. First we have a look at how economic-based theories, such as voluntary disclosure theory and signalling theory could contribute to understand AI voluntary disclosure as both were previously widely applied to corporate disclosure in general (Lu and Wang, 2021). These economic-based theories consider investors and shareholders the centre of interest. On the other hand, legitimacy theory, which is a socio-political theory, emphasise the dialogue between a company and society including all stakeholders (Suchman, 1995).

Starting with economic-based theories, voluntary disclosure theory has some connections with game theory and states that companies only disclose favourable information voluntarily, leaving unfavourable facts unreported (Dye, 2001). This theory was initially used to explain anomalies related to financial information. However, recently, it has been expanded to non-financial reporting as both financial and non-financial aspects are relevant for investors to make informed decisions (Zhou et al., 2017).

Thus, voluntary disclosure theory explains that companies tend to overcome a mandatory level of disclosure to appeal to their shareholders and potential investors, however, they also tend to disclose only favorable information to obtain positive economic effects. Therefore, as AI disclosure is completely voluntary character and there are no guidelines on what and how to report, some companies might be motivated to disclose information about their AI projects and products to appeal to their shareholders as a smart, digital-savvy and innovative company.

Signalling theory is another economic-based theory. This theory is based on a concept of reducing information asymmetry (Connelly et al. 2011) between a signaller (company) and receivers (stakeholders). A signal in this context is understood as a signaller's future strategies with the main aim to attract investors attention by highlighting positive aspects which are hidden to external subjects (Lu and Wang, 2021). In this context, voluntary AI disclosure can have a positive impact on investors' perception of a company's value and can lead to decisions that would positively affect the company.

Several studies found that legitimacy theory can provide a satisfactory explanation of why companies voluntarily disclose non-financial information (Bonsón and Bednárová, 2013; Campbell et al., 2003; Deegan, 2002). Thus, considering CDR to be a part of non-financial disclosure, legitimacy theory might offer a reasonable justification for voluntary AI disclosure as well (Bonsón et al. 2021b). Corporate legitimacy is based on an assumption that all the actions of a company are in compliance with societal norms, values and expectations (Suchman, 1995). Thus, companies try to legitimize their actions and practices to gain acceptance from different stakeholders, not only shareholders (Deegan, 2002). It has been demonstrated that higher transparency helps to increase trust in a company. Nevertheless, to gain legitimacy via transparency means that companies disclose relevant information related to the most material issues or concerns of society. Hence, materiality of disclosure is what matters when it comes to the connection between transparency and legitimacy (Bonsón and Bednárová, 2022). Nowadays, it is AI technology that raises certain concerns, in particular black box ADMs or higly sensitive applications such as facial recognition (Osburg, 2017; Thorun, 2018).

2.5.4   ADM Disclosure

A summary of our recent research on ADM disclosure with more details regarding the objectives, research questions, methods, findings, and conclusions is provided below (Bonsón et al. 2022a).

2.5.4.1   Objectives on ADM Disclosure

As the rapid proliferation of AI tools in our daily lives continues to increase, there are legitimate worries in society, which call for the ethical and responsible development and use of these tools. Therefore, our study aims to shed some light on whether companies are being transparent about the way they use AI. As a result, corporate reports with ADM-related mentions were analysed, as these applications imply the highest concerns because the decision-making is left to machines.

2.5.4.2   Research Questions

The research questions focused on three main aspects 1) Do Western European companies disclose information about the ADM use in their corporate reports? 2) What content type they disclose? 3) Which factors determine ADM disclosure.

2.5.4.3   Methods of Analysis on ADM Disclosure

As CDR appears to be a new layer of CSR, we can expect adjustments in corporate reporting in the upcoming years too. The first adopters of CDR disclosure would very likely be large corporations. Pressure from their stakeholders to be more transparent about their digital responsiblity is higher than for small and medium-sized companies due to higher potential impact on society in case of irresponsible actions. Consequently, large companies might try to legitimize their usage of AI via voluntary disclosure.

In our research we focused on a sample of large Western European companies and analysed one aspect of CDR, the ADM disclosure.

Sample

For the analysis, both annual and sustainability reports (for the years 2018 and 2019) of large listed Western European companies (Austria-ATX 20; Belgium-BEL 20; Denmark-OMXC 25; Finland-OMXH 25; France-CAC 40; Germany-DAX 30, Greece-FTSE 20, Ireland-ISEQ 20, Italy-INVIT 40; Netherlands-AEX 25; Portugal-PSI 20; Spain-IBEX 35, Sweden-OMXS 30) operating in 11 sectors (Energy; Materials; Industrial; Consumer discretionary; Consumer staples; Healthcare; Financial; Information technology; Communication services; Utilities; Real estate) were collected. In 2018, 474 and in 2019, 488 PDF documents (annual and sustainability reports) were downloaded making a total of 962 documents for our analysis

Content Analysis

Content analysis is a research method which enables us to investigate a wide spectrum of problems by systematically and objectively identifying specific characteristics of an analysed content of any type of communication (Holsti, 1969). Thus, it is considered to be a standard method for systematically comparing the content of communications such as advertisements, media stories, websites, etc. (Kolbe and Burnett, 1991; Smith et al., 2012). As corporate reports are company communication to their stakeholders, in our study we applied the content analysis on annual and sustainability reports of the analysed companies. The traditional content analysis consisted of manual process of sampling, coding, analysing and consolidating the results. In recent years, software such as R (R Core Team, 2018) emerged, which allows us to make the content analysis process more automated. In the following subsection, the automated content analysis via bigram automated scraping is described in more details.

Automated content analysis: Bigram automated scraping

Text-mining was conducted using the R software (R Core Team, 2018) and the text was extracted through "extract_text()" function of the "tabulizer" R package, with the ability to convert a PDF file into a single character vector (1x1). After some necessary adjustments (sentence separation and text cleaning), a dictionary of bigrams was proposed including keywords such as algorithm (s) (ic); automat (ed) (able) (ion) (ic) and associated words such as decision (s). Based on this, an automated content analysis was performed (Bonsón et al., 2022a). The whole process with more details is depicted in Figure 7.

Statistical analysis

The dependent variable was a dummy variable indicating whether or not the report mentions ADM. Independent variables were geographical area, sector, company size and a sustainability leadership. The generalized linear model (logit regression) was applied to analyse the data.

$$\{logit(ADM) = \beta 0 + \beta 1 GeoArea + \beta 2 Sect + \beta 3 Size + \beta 4 SustaiLead\}$$

Figure 7   Flowchart of pdf text mining

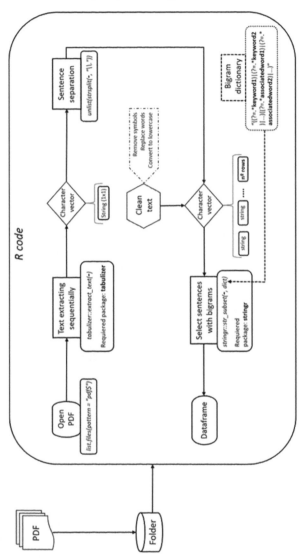

Source: Bonsón et al. (2022a)

2.5.4.4   Findings and Conclusions on ADM Disclosure

The findings showed that only a few companies report on ADM in their corporate reports. Out of 962 analysed documents, there were only 28 mentions of ADM, which were associated with 20 companies. Another finding that emerged from our research was that ADM reporting is related to the sector in which the company operates and that companies operating in the financial sector are more likely to disclose such information.

Regarding content type, the most commonly disclosed topics were: ADM for credit risk assessment; ADM responsibility; medical and diagnostic algorithms; and other. The category other was mostly related to the electronic sector talking about AI used in different types of electronic devices such as smartphones, smartwatches, home robots, etc. Although ADM for credit risk assessment was the most frequently disclosed topic, the level of detail was rather vague, e.g. there was a lack of details on which processes in credit risk assessment are fully automated and which are just assisted by the AI. ADM responsibility was the second most frequently disclosed topic, but it presents the same weaknesses as credit risk assessment. Some companies only express general concerns regarding uncontrolled use of AI. Other companies provided more specific disclosure related to avoidance of bias and discrimination; digital rights; personal data processing, etc. (Bonsón et al., 2022a).

2.5.5   AI Disclosure Framework

As we have seen in the previous chapter, only a few companies disclose information about AI in their corporate reports. Those were mostly companies operating in financial and telecommunication sector. The way those companies report on AI vary from company to company in terms of extent of disclosure, content type, level of details provided, etc. The reason for that is that AI reporting has a voluntary character and there are no guidelines to help companies decide what and how to report. However, recent initiatives such as the regulation proposal for AI discussed previously, are expected to increase interest in AI transparency in the upcoming years.

2.5.5.1 Objectives

The main objective of our next study was to seek consensus applying a multi-stakeholder approach and propose a set of elements to structure information related to AI disclosure. That is to say, to provide a framework to companies willing to disclose this kind of information and guidelines on what and how to report.

2.5.5.2 Methods: Focus Groups and Questionnaire

Disclosure elements proposal

An initial set of elements to be disclosed was proposed based on the EC's White paper on AI – A European approach to excellence and trust (2020), Artificial Intelligence Act (2021) and Digital Rights Charther – XXIII Digital rights related to AI (SEDIA, 2021). This initial set of elements was refined during the focus groups with BIDA Observatory and the New Technologies and Accounting Commission of Spanish Accounting Association (AECA). The collaboration between BIDA Observatory and New Technologies and Accounting Commision of AECA represents a multi-stakeholder approach as it combines regulatory bodies, public organizations, academics and business world.

The refined proposal was later validated by the international community of academics, accountants, auditors, managers, and others through an online questionnaire.

Elements validation via questionnaire

The objective of the online questionnaire was to validate the relevance of each of the elements proposed. The respondents were asked to rank each item on the scale from 1 (least relevant) to 5 (most relevant) and add any other element they would consider relevant.

The questionnaire survey was conducted from January to March 2021 with a total of 44 responses which translates to a 58,6% response rate. Figure 8 depicts the professional profile of the respondents.

Figure 8 Respondent profile

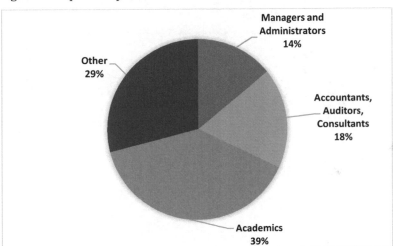

Source: Author's elaboration

*2.5.5.3   Findings and Conclusions –   ADM Disclosure Framework
            Proposal*

The disclosure elements consists of two main categories: general and specific (related to ADM due to its elevated inherent risk). The category general is further broken down into three subcategories: AI governance
model (including 3 reporting items); Ethics and responsibility (including 3
reporting items); Strategy (including 4 reporting items). The specific
ADM category includes 3 reporting items. The validation of proposed elements through the online questionnaire showed that all items were validated (obtaining 4 points or more on the scale from 1 to 5, each). Based
on this validation, an AI disclosure framework consisting of two parts:
general elements and specific ADM disclosure was developed (Tables 2
and 3).

Table 2 AI disclosure framework proposal for general elements

General elements	
AI governance model	(GM1) Governance structure, roles, responsibilities, departments and committees
	(GM2) Continuous monitoring, control and internal verification
	(GM3) External verification systems or processes (of high sensitive applications)
Ethics and responsibility	(ER1) AI principles followed by the company (own, EU, IEEE, OECD, etc.),
	(ER2) Relevant ethical issues, especially in relation to sensitive applications such as facial recognition, black box ADMs and recommendation systems
	(ER3) Mechanisms for users to receive meaningful explanations
Strategy	(S1) AI risk and impact analysis
	(S2) Relations with stakeholders, participation in organizations and forums related to the responsible development of AI
	(S3) Training activities (internal and external) on responsible use of AI
	(S4) Projects focused on meeting the Sustainable Development Goals (SDG)

Source: Bonsón and Bednárová (2022)

AI governance model

Regarding AI governance model, companies should provide information related to the AI governance structure, existence of AI departments, committees, and AI-specific roles such as data protection officer, explain how continuous monitoring of AI processes is performed, and if any external verification systems and processes take place.

Ethics and responsibility

In this section, companies might disclose if they follow any international principles related to AI ethical usage such as IEEE, UE, OECD or similar, or if they have developed their own AI code of ethics. Additionally, they should disclose any issues related to privacy matters, or the application of highly sensitive apps such as facial recognition or ADM. Similarly, a company could provide information about the existing mechanism for advice or concerns related to responsible use of AI.

Strategy

In terms of AI strategy disclosure, most atttention should be paid to how AI related risks are managed. Thus, risk assessment indicating why some applications imply higher risks than others and how they are identified and classified within a company should be clearly explained. Additionally, in this section, a company can provide information about activities related to responsible development and the use of AI such as training activities, participating in forums, or projects focused on SDGs.

Table 3 AI disclosure framework proposal for specific ADM elements

ADM disclosure	
Automatic decision-making system (ADM) transparency	(ADM1) System description: type, input data, algorithm design, human supervision, etc.
	(ADM2) Statement that the system complies with the digital rights regulation related to AI or, if applicable, that algorithmic non-discrimination, transparency, auditability and explicability of the algorithm are ensured, as well as its accessibility, usability and reliability.
	(ADM3) Verification of compliance by independent agents.

Source: Bonsón and Bednárová (2022)

ADM disclosure

After risk assessment, the applications with highest risk would pass to the category that requires more specific disclosure. This requirement is in compliance with EC's approach, which distinguishes four levels of AI risk categories with the aim to protect individuals against negative implycations of AI applications, but do not cause unnecessary burden for companies which are not using very invasive AI technology. Thus, special attention should be paid to black box ADM as the decision-making is left to the algorithms and as those were developed by humans, are neither error nor bias-free. That is why, it is necessary to know how they were built; how the data are collected, stored, and managed; how the algorithm works, etc. Similarly, biases and mitigation plans should be disclosed among other aspects such as human supervision, auditability, explicability, accessibility, usability and reliability of the algorithm, as well as external verification.

2.6 Conclusions

When companies use AI in their processes, they should be transparent about why and how the AI technology is being used, particularly if it involves highly autonomous AI tools such as ADMs. As such disclosure is a new trend, there is a lack of guidelines to help companies with this reporting.

With this effort we wanted to draw attention and increase awareness of an urgent need for internationally recognized and generally accepted AI reporting standards, the same as we have for financial reporting. Such standards should represent global consensus and best practice for publicly reporting the impacts of company AI usage. AI disclosure based on the standards would also help guarantee a company's commitment to fundamental rights such as freedoms, privacy, data protection and non-discrimination and would help to shape its corporate digital responsibility (Bonsón and Bednárová, 2022). We believe that our study might have contributed to this field by proposing and validating a set of elements related to AI and ADM disclosure, shaping the future framework or standards in this area.

2.7 Future Research and Limitations

Nevertheless, despite the fact that our study is one of the first attempts to standardize AI disclosure, due to the fast evolution of AI technology, a continuous assessment of disclosure elements should be conducted. Similarly, regarding the research on ADM disclosure, more extensive research is still needed to explore the current practices in AI disclosure in general (not only ADM) and the determinant factors of such disclosure.

3 Continuous Reporting

This chapter deals with a new way of corporate disclosure – continuous reporting.

3.1 Background

The main market forces that are driving changes toward continuous reporting are: 1) data is produced almost continuously, 2) stakeholders expect up-to-date information and transparency on a variety of issues, 3) technological advances (XBRL, blockchain, software processing Big Data, artificial intelligence (AI), and Internet of Things (IoT)).

Current accounting processes appear obsolete and there is a need to create a reporting format that would cover modern stakeholder information needs. Although current legislation in most jurisdictions has moved toward sustainability reporting requiring enhanced disclosure and transparency from large corporations in terms of their environmental and social impact as well as financial performance, current reporting practices are far from being efficient. Annual reports or integrated reports, which sometimes replace traditional annual reports, might be covering different stakeholder information needs but lack efficiency in the way the information is collected, processed, and presented. These reports are still long and getting longer. The format of reporting should keep pace with new challenges and opportunities such as technological advances which are transforming the business environment. In the same way, the premise of "doing well by doing good" transformed the content of corporate reports, the new premise formulated as "do not inform me more, but inform me better" could transform the way information is presented. Hence, this chapter focuses on the implications of the synergy of xbrl, blockchain, Big Data, artificial intelligence and IoT on the continuous reporting. The aim of my study is twofold. First, I aim to analyze the strengths and shortcomings of every component of this synergy. Second, I aim to outline a semi-automated continuous reporting model where IoT feels, AI thinks, BC remembers and XBRL compares.

3.2 Literature Review

3.2.1 *Big data , Artificial Intelligence and Blockchain in Continuous Accounting*

Moffitt and Vasarhelyi (2013) and Vasarhelyi et al. (2015) analyzed the impact of Big Data on accounting and auditing. The studies discuss an overall framework of Big Data in accounting. They point out the changing nature of accounting records and the integration of non-traditional data into accounting and auditing and the new opportunities and challenges for audit analytics enabled by Big Data.

According to Sgantzos and Grigg (2019), BC as an inmutable storage medium might become a prominent platform for AI. Similarly, Stein (2018) examined current market forces linked to technology and the increased influence of stakeholders on the reporting process. He analyzed the implications of blockchain and artificial inteligence on financial continuous reporting.

Corea (2019) and Marwala and Xing (2018) analyzed how blockchain implementation can be enhanced by applying AI solutions, pointing out that the synergy of AI and blockchain can lead to numerous possibilities, including but not limited to continuous reporting.

3.3 Theoretical Background: Stakeholder Theory

Stein (2018) points out two main shortcomings of current reporting, the narrow applicability of information presented which is focused mostly on financial shareholders, and the time lag which implies that reported information is often a couple of months out-of-date.

Different stakeholder groups, either on an individual or institutional basis, are increasingly interested in information beyond the financial performance of the company. Thus, regulators, NGOs, governmental bodies, and also consumer groups press companies to provide a broader view of the information and require the insights on how the results of the company were achieved in terms of environmental and social impact.

According to Drew (2017), the advances in technology such as automation and digitization of corporate information benefit all types of stakeholders, not only shareholders. O'Brien (2016) also argues that technological innovations, such as IoT, improve corporate disclosure by increasing the operational efficiency and providing more comprehensive reporting that satisfies all stakeholder groups. Stakeholder reporting and communication of different types of information to a variety of stakeholders is the responsibility of management teams and eventually leads to improved engagement (Stein, 2018).

Hence, technology-based continuous reporting covering not only financial but also non-financial aspects of business seems to be a logical step forward. This will satisfy different stakeholder groups increasingly demanding more information on a more regular basis.

3.4   Integrated Navigation Chart

In this section, I elaborate on the idea of integrated reporting navigation chart initially elaborated during my PhD studies, which provided some practical recommendations for further development of CSR reporting (Figure 9). This figure depicts the shift from the traditional reporting model to the new forms of reporting where Integrated Reporting is the latest trend.

The Integrated Reporting Framework proposes connecting the corporate report with company social networks and other digital objects. I applied AECA's[2] Integrated Scorecard Key Performance Indicators (IS-KPIs) due to its wide coverage and well-defined KPIs for ESG disclosure. In my proposal, these connections can be achieved by an integrated reporting navigation chart, which would combine XBRL elements by providing hyperlinks to all related digital objects such as PDF files containing financial and sustainability reports, YouTube videos, and social media channels providing a wider context to reported information. This way, the standardized XBRL data could effectively connect narrative explanations regarding financial, environmental, social, and governance performance with KPIs (quantitative data). Thus, applying the embedded XBRL taxonomy would enable better comparison and interchange of corporate data as well as seeing the evolution and possible impacts of reported KPIs in a wider context of sustainability. Having only quantitative KPIs does not completely explain the business model of the company and how corporate strategy affects corporate performance and corporate value. The whole framework would be based on the reporting principles presented by the IIRC: strategic focus, materiality and reliability, responsiveness and stakeholder inclusiveness, future orientation, and connectivity of information. Hence, the navigation chart would allow the connection of information into a coherent and integrated whole, which is one of the most important attributes of integrated reporting.

---

2   Asociación Española de Contabilidad y Administración de Empresas (AECA)

Figure 9   Integrated navigation chart

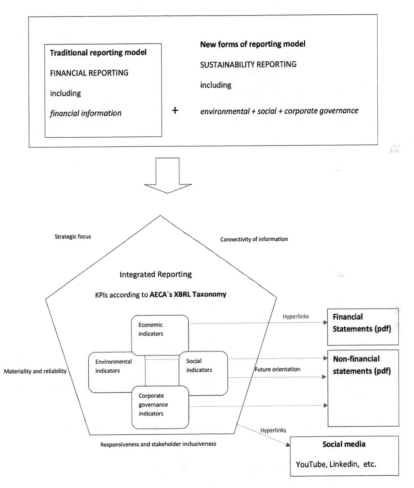

Source: Author's elaboration

3.5   Principles Continuous Reporting Model

Nevertheless, as technology has advanced, the proposal of integrated navigation chart based only on XBRL and hyperlinks also becomes obsolete as new technologies enable efficient methods of data collection, process, and storage. Thus, this chapter aims to propose a reporting model that might be continuous and semi-automated by taking advantage of the synergy of XBRL, blockchain, Big Data, artificial intelligence and IoT, a model **where IoT feels, AI thinks, BC remembers and XBRL compares.**

As the principles presented by IIRC are rather abstract with some level of overlapping, based on comprehensive literature and technological aspects analysis, and desk research, the following principles for creation of a continuous reporting model have been proposed.

Principles:

- Materiality
- Connectivity (Financial, ESG, CDR disclosure)
- Comparability (XBRL-based)
- Traceability (Blockchain-based)
- Automation (IoT, Smart Contracts)

Materiality principle implies that instead of adding more information and making already long annual reports even longer, only relevant information should be disclosed. Nevertheless, a company should take into consideration the information needs of all stakeholders, not only shareholders. Therefore, it should go beyond financial information and even ESG disclosure as nowadays, due to ongoing digital disruption, the CDR becomes a concern due to its possible ethical implications.

Connectivity principle is about combining a) contentwise financial data with non-financial disclosure (including CSR and CDR) and b) formatwise numerical (KPIs) and narrative information (block tags).

Comparability principle emphasizes the importance of disclosing information in a format that would allow an easy comparison with other companies. As XBRL reached its maturity and is currently a disclosure format for large corporations, it would play an important role in current and future corporate disclosure. Regarding financial disclosure, there is an existing taxonomy. As for non-financial disclosure, the task is slightly more complicated, but there is an array of initiatives on a national and international level aiming to develop and agree upon an internationally recognized and generally accepted XBRL ESG taxonomy as well. Regarding CDR disclosure, one of the possible approaches to disclose CDR information in the XBRL format would be to use "block tagging" as it might be difficult to develop KPIs for CDR reporting. This approach has already been approved for Notes and listed European companies should put them into XBRL format from 2022.

Traceability principle points out the importance of using a platform that would guarantee traceability of information to check data veracity for auditing, etc. Thus, blockchain as a database based on DLT might play an important role.

Automation principle is based on incorporating IoT and smart contracts into corporate data collection, storing, and reporting. IoT sensors incorporated into smart contracts would guarantee a continous update of information for management, financial and non-financial accounting.

These principles are not meant to replace the principles of integrated reporting presented by IIRC (Table 4), but they should instead complement them. Given the technological nature of continuous reporting, new principles related to technological aspects have been added.

Table 4   Integrated reporting principles vs. continuous reporting
principles

Integrated Reporting Principles	Continuous Reporting Principles (proposed)	Character
Strategic focus	Materiality	Non-technological
Materiality and reliability		
Responsiveness and stakeholder inclusiveness		
Future orientation		
Connectivity	Connectivity	Non-technological, Technological
-	Comparability	Technological
-	Automation	Technological
-	Traceability	Technological

Source: Author's elaboration

In the following subsections, the arguments for including the principles of comparability, automation and traceability will be outlined.

3.5.1   Comparability Enabled by XBRL

3.5.1.1   Regulation and initiatives

Comparability is, and will continue to be, a driving force. Indeed, regulators across the world have been introducing legislation requiring companies to submit their corporate reports in digital format. The structured disclosure requirements in the US have mandated larger US-based listed companies to submit digital corporate reports since 2009 (SEC, 2009). In

Europe, until now, entities have been required to file annual financial reports in PDF format, which has several frailties e.g. difficulty in comparing the performance of various companies efficiently. Thus, the primary reason behind the introduction of the electronic format was to boost transparency and enable more efficient analysis of such information for investors, authorities, and other users of financial statements.

EU efforts for the harmonisation of transparency dates back to teh Transparency Directive (2013). The ESMA' mandate on the ESEF is contained in the revised Transparency Directive. Article 4 (7) states that the ESMA shall develop draft regulatory technical standards to specify the electronic reporting format, pointing out that a harmonised electronic format would be very beneficial to different stakeholders, since it would make reporting easier and facilitates accessibility, analysis and comparability of financial statements. A specific mention of using XBRL was made. Transparency Directive refers to Directive 2013/50/EU of the European Parliament and of the Council of 22 October 2013, amending Directive 2004/109/EC of the European Parliament and of the Council on the harmonisation of transparency requirements in relation to information about issuers whose securities are admitted to trading on a regulated market, Directive 2003/71/EC of the European Parliament and of the Council on the prospectus to be published when securities are offered to the public or admitted to trading and Commission Directive 2007/14/EC laying down detailed rules for the implementation of certain provisions of Directive 2004/109/EC Text with EEA relevance (Eur-lex, 2013).

From the 1st of January 2020, all companies listed in the European Union public markets are obligated to mark up their annual consolidated financial statements with XBRL tags and report them to the local regulator using the electronic format. This is based on the European Single Electronic Format (ESEF) mandate issued by the regulatory authority of the EU, the ESMA. It relates to primary financial statements such as Balance Sheet, Income Statement, Cash Flow Statement and Statement of Changes in Equity. Nevertheless, from 2022 block-tagging of Notes will also be mandatory. This is in compliance with Transparency Directive and based on the Commission Delegated Rregulation (EU) 2019/815, which is supplementing Directive 2004/109EC of the European Parliament and of the Council with regard to regulatory technical standards on the specification of a single electronic reporting format. Consequently, the European Securities and Makets Authority (ESMA) was assigned the responsibility of developing regulatory technical standards (RTS) to specify this electronic reporting format to make reporting easier for issuers and to facilitate accessibility, analysis and comparability of financial reports. Thus, the ESMA provided guidance in the form of the ESEF Reporting Manual.

The summary of ESEF requirements is provided below:

- All Annual Financial Reports (AFRs) shall be prepared in xHTML, which is human-readable and can be opened with any standard web browser;
- AFRs containing consolidated IFRS financial statements shall be labelled (marked up) by using XBRL 'tags', which make the labelled disclosures structured and machine-readable;
- The XBRL 'tags' shall be embedded in the xHTML document using the Inline XBRL technology, which allows the benefits of XBRL tagged data to be combined with the human readable presentation of AFRs. This way, companies would create an iXBRL version of their annual report (ESMA, 2019).

The technological elements of ESEF are explained below:

- xHTML – Refers to extensible hypertext markup language. It belongs to the family of XML markup languages and extended version of common HTML (hypertext markup language), which is a language in which web sites are formulated. Thus, it is human-readable and can be displayed in standard internet browsers.
- XBRL – Stands for extensible business reporting language and it is an open international standard for digital business reporting managed by a global non-profit consortium XBRL International. It allows software supported analysis and it is already used in a number of different juris-

dictions. Currently, it is considered the only appropriate markup language for financial statements.

- iXBRL – Refers to Inline technology and is being used to embed XBRL tags in the xHTML document.

Initially, all publicly trated securities issuers in the EU (and the EEA – European Economic Area) shall prepare annual financial reports in ESEF for the financial year starting on 1st January 2020. Nevertheless, later the EC provided the option to postpone the mandate to the financial year beginning on or after January 2021, enabling voluntary submission until then. Indeed, many companies have presented their anual reports in iXBRL format for the accounting year 2020, but there are also many which decided to prepare better and postpone it.

3.5.1.2   New Reporting Process

In this section, a description of the new process in the preparation of annual company reports is outlined. Firstly, one additional step should be added to the traditional reporting process, which would consist of mapping financial statement items to the ESEF taxonomy. Secondly, a tool or service to generate financial statements in XBRL format should be implemented. There are an increasing number of companies providing software solutions and advisory services in the process of preparing financial statements in XBRL formats. The creators of applications are teams of expert accountants and IT specialists.

The new format enables a smart aggregation of reporting data and makes them accesible to investors, analysts, regulators for comprehensive analysis but it has more than just technical aspects. Communication and training plays an important role. Therefore, staff should be provided with appropriate training on ESEF requirements and stakeholders such as investors, analysts, etc. should be informed about the changes in reporting format.

Steps to prepare an inline XBRL report (ESMA, 2019):

1) Issuers need to familiarise themselves with the requirements set out in the **RTS on ESEF**; **the ESEF Reporting Manual** and **ESEF taxonomy**. Taxonomy is essentially a dictionary of accounting terms. ESEF taxonomy builds on IFRS Taxonomy. IFRS Taxonomy is based on IFRS Standards and common reporting practices, which are elements not specifically mentioned in IFRS Standards but are consistent with them. Those are elements that have been frequently disclosed across a wide range of companies across juristictions.

2) Preparation of a correlation table, which means mapping IFRS financial statements to ESEF taxonomy. In this step, for each item of the IFRS based financial statement, the ESEF taxonomy element which has the closest accounting meaning should be marked up. This process is called tagging and it means attributing the most appropriate element chosen from an ESEF taxonomy to financial data.

3) If there are lines which do not have a corresponding element in ESEF taxonomy because they are specific to an issuer or industry, an extension should be created and anchored. The extension elements allow issuers to tag information which is unique for a company, but for comparative purposes it will be anchored to an existing taxonomy element. Anchoring means to create a link through an XBRL relationship.

4) Use Inline XBRL technology to embed XBRL tags to xHTML (human-readable) format.

5) Submission of the ESEF reports. Companies are obligated to submit the ESEF reports to a competent national authority, not to ESMA.

3.5.1.3   *ESEF Reports Repository*

Until now, there has been no easy way to access financial statements from across Europe. There is an announcement of the creation of the European Single Access Point (ESAP), which will provide a free single point of access to information on EU-listed companies and investment products, something like EDGAR in the US or EDINET in Japan. ESAP is scheduled for launch in 2024 (XBRL, 2021).

In addition, XBRL International built a repository[3], which currently includes over 700 ESEF Annual reports. Although many countries chose to delay the mandate due to the Covid-19 pandemic, a small number did go ahead with mandatory reporting and some companies chose voluntary ESEF reporting to test out the reporting process.

3.5.1.4 *The Role of XBRL in Continuous Reporting*

The financial KPIs can be tagged directly to existing ESEF taxonomy elements or via anchoring the extension. For the narrative disclosure, like Notes, the block-tagging of text is proposed. Thus, a similar approach could be adopted for the narrative disclosure on CSR or CDR in the future. It should also be highlighted that RTS allows, on a voluntary basis, tagging images, data in graphics, etc. Therefore, the possibilities that XBRL reporting offers could provide end-users with additional meaningful information for their analyses.

In addition, the comparability that XBRL offers can play a crucial role in the continuous reporting model. New technologies, such as IoT, AI, and blockchain open many doors for continuous reporting. Nevertheless, if data collection and its process become scattered due to a variety of technological elements that would be able to collect, store and report a big amount of unstructured data, comparability will be threatened, which would complicate the task for analysts. Therefore, in the model of continuous reporting, where IoT feels, AI thinks, BC remembers and XBRL compares, the participation of XBRL remains crucial.

---

3   filings.xbrl.org

3.5.2   Automation Enabled by IoT and Smart Contracts

The continuous reporting framework I propose is based on a combination of traditional data (manually acquired) and a variety of data obtained via IoT (scanners, sensors), oracles, etc.

The IoT collects and analyzes a huge amount of data via different sensors. Its key components are sensors, processors and communication networks.

Traditionally, the measurement of many accounting items depended, and still depends, on cost/benefit considerations. Thus, many aspects are left to estimation, e.g. inventory consumption methods such as FIFO/LIFO can be replaced by exact consumption detection via e.g. radio frequency identification (RFID). Nevertheless, with the evolution of new technologies, e.g. IoT, economic transactions can be automated in most of the steps (if not all) of the process: identifying, measuring, recording and reporting. The possibility of measurring different non-financial indicators via sensors (e.g. CO_2 emissions, water consumption, etc.) allows companies to conduct sustainaibility management and reporting more efficiently too. IoT, if combined with machine learning, can be truly transformative for the manufacturing industry in particular. Managers could have a remote view of every machine in the factory, so they can see the actual performance data of each piece of equipment. In addition, machine learning-powered proactive maintenance can even anticipate a possible fault in a machine so replacement parts could be preordered, and a delay in the production process would be minimized.

Smart contract is an automatically executing computer program and it can perform various operations such as information storage, funds transfer, etc. It is designed the following way: when the program recognizes that conditions agreed in contract are met, the transaction is automatically executed and recorded on blockchain, which guarantees immutability of the records. Thus, automation via smart contracts can lead to higher efficiency of supplier payments and clients collections management or VAT payments, for example. Smart contracts can be connected to IoT, so when a sensor detects that a material supply reaches the warehouse, the inventory is automatically registered in accounting and payment to the supplier is released as programmed (e.g. 14 days after, or after reaching a certain cash balance).

3.5.3   Traceability Enabled by Blockchain

This feature received more attention with the evolution of blockchain, a traceable and immutable database. As explained previously, blockchain is a decentralized distributed ledger that allows peer-to-peer transactions secured by cryptographic algorithms and consensus mechanisms. Decentralized implies that no central authority is needed to overlook the transactions and ledger, a registry of transactions, is maintained simultaneously by all the participating nodes. In addition, transactions between peers use algorithms that cryptographically protect transactions to avoid any tampering. Consensus mechanism, on the other hand, allows the ledger that is distributed to be considered "a unique source of truth". Digital token is anything that represents a value or a title of ownership and can be exchanged between parties. For example, Bitcoin is a cryptocurrency that runs on blockchain and cryptocurrency is a digital token exchanged on blockchain using cryptographic algorithms to secure peer-to-peer transaction. Thus, cryptocurrency is a type of a digital token that represents a monetary value, but digital token is a much wider term. It might be a document, pallet of goods, etc. Regarding the blockchain protocol, we distinguish between two main types: permisionless and permissioned blockchain (although there might be some adjustments and variations of both). Permissionless blockchain is a blockchain protocol that allows anyone to join the network. Permissioned blockchain, on the other hand, is a blockchain protocol that requires authorization to join the network. That is why, enterprise blockchain would very likely be permissioned between nodes like suppliers, clients, partners, etc. that know each other.

There already are developers of enterprise solutions where customers explain their business logic and needs and the rest is taken care of . BaaS (Blockchain as a Service) providers include IBM, Oracle, AWS, Accenture, Microsoft, Amazon, and others. The most well-known enterprise blockchain platforms are: Fabric (Hyperledger, IBM), Corda (R3 consortium), EEA (Enterprise Ethereum Alliance), Quorum (spinoff of Ethereum, J.P. Morgan), or Ripple.

3.5.3.1   Blockchain Role in Continuous Reporting

One of the most important roles of blockchain in continuous reporting would be that when data are obtained continuously, no third party would be necessary to continuously check the veracity of transactions. Something that would not be feasible without blockchain. Thus, it enables kind of triple entry bookkeeping where the third party is blockchain itself. All records are cryptographically sealed, registered and distributed among all the nodes of the network with its consequent time stamp. Therefore, we have certainty that information has not been modified and every transaction can be traced. In addition, automation of certain data collection can be done through smart contracts even with combination of IoTs. Thus, actions like ordering payments when certain conditions are met, giving alerts on inventory status/or automatically ordering inventories, issuing invoices, warnings about treasury balances, and many others might be automated. This way, blockchain can serve as a database for all the transactions of a company, although adjusting the visibility of certain data to certain nodes can be programmed too.

3.6   Semi-automated Continuous Reporting Model

Figure 10 depicts a simplified schema of a semi-automated continuous reporting model (SACRM), which would be based on the previously discussed principles.

As can be seen in Figure 10, SACRM would follow the principles of materiality, connectivity, comparability, automation, and traceability. Thus, principles of IR have been extended by technologically based principles (comparability via XBRL, automation via AI, and traceability via blockchain). Regarding the content, financial reporting and ESG approach is complemented by CDR disclosure, the importance of which has been

Figure 10 Semi-automated Continuous Reporting Model

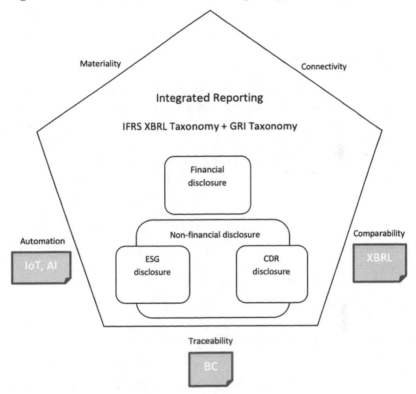

Source: Author's elaboration

discussed in chapter 2. The proposed basis of financial reporting is IFRS XBRL taxonomy (as indicated in ESEF mandate). It is expected that an internationally recognized and generally accepted XBRL taxonomy for non-financial indicators will soon appear, which can be based on widely known GRI Standards or other relevant Standards. Narrative non-financial reporting can use block-tagging as it is already recommended for Notes.

3.7   A Roadmap of the Development Process

Figure 11 depicts a roadmap to develop SACRM. Initial research should be based on the validation of proposed principles as well as an analysis of technological functionalities of continuous reporting. After an initial period of research, a consultation paper can be published including a cost-benefit analysis. This would be followed by a review of responses and publication of a feedback statement including a second cost-benefit analysis. After that, relevant technical specifications could be developed and tested through field tests with the companies signed up to the pilot project. The objective of the pilot project will be to assess through practical experience if, and to what extent, the specifications should improve in order to increase usefulness of the continuous reporting model and reduce the burden on preparers. The lessons learned will be incorporated in the Draft Continuous Reporting Model, which can serve as guidance for the continuous reporting framework. This draft will undergo further scrutiny from different stakeholders (including issuers and users) who would be invited to provide complementary feedback and recommendations for improvements.

Figure 11   An overview of the development process

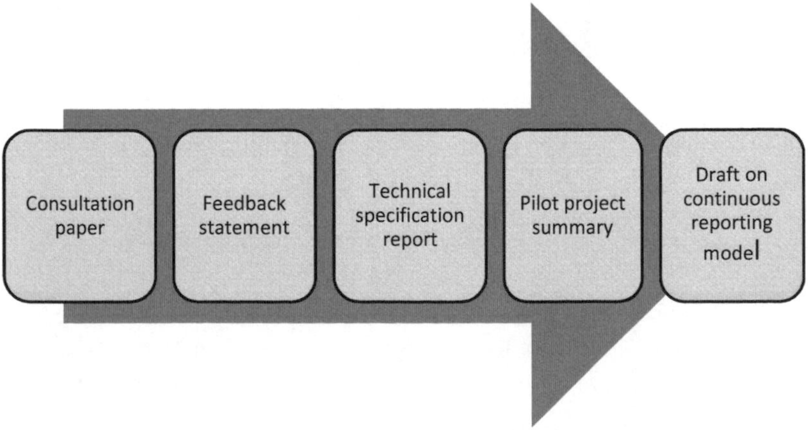

Source: Author's elaboration

3.8 Challenges

Dai and Vasarhelyi (2017) introduced three different contexts of challenges for blockchain in auditing: technological, organizational and environmental. Regarding the continuous reporting model based on IoT, AI, BC and XBRL the following contexts of challenges are identified:

Technological and economic context: refers to the technical complexity which requires financial and time resources. As Dai and Vasarhelyi (2017) pointed out, a company might struggle to find business partners to share the blockchain ledger with, for example. Thus, the challenge here is to build a functional infrastructure and minimize the cost and complexity of AI, IoT, and BC implementation. Integration of blockchain with an existing systems of records is one of the significant challenges. In addition, when programming smart contracts, concensus from all participants is needed.

Organizational culture context: refers to willingness of managers to welcome innovation (benefits must exceed the potential costs) and a resistance to change within a company's culture as well as a level of employees' adaptability and willingness/capacity to acquire new skills. Thus, upskilling will play a crucial role in upcoming years.

Regulatory context: points out the essential role of regulators in adoption of new technologies within the accounting and auditing ecosystem. Regulators are supposed to help govern different fields. Therefore, a general understanding of the main implications and limitations of new technologies such as blockchain is necessary for them to make informed decisions.

General public context: Last but not least, innovation can be indirectly impeded if there is a lack of technological acceptance by the general public. Conservative stakeholders might be risk-adverse. Therefore, a certain level of education among the general public would be essential.

3.9 Conclusions

The main advantages of moving towards continuous reporting based on technologies such as AI, IoT, blockchain and XBRL would be: up-to-date and accurate information, less human errors, transaction security and easy comparability, which would ultimately lead to higher efficiency for analysis and decision making.

Thus, by incorporating IoT, real-time data will be available (instead of quarterly or annual data presentation) and automation would decrease human errors and replace estimations with accurate data. Blockchain would guarantee the security of transactions and XBRL the comparability.

Principles for continuous reporting such as materiality, connectivity, comparability, automation, and traceability are based on the literature and technology review, and desk research. Nevertheless, they could be validated by a multi-stakeholder approach, similar to what has been done with the AI disclosure framework proposal.

Even though AI principles and guidelines have been criticised for beeing too abstract, they certainly laid foundations for an AI regulation proposal. Thus, the proposed continuous reporting model, if it gains acceptance, can start a debate on, or serve as a basis for, future standardization-regulation purposes in terms of continuous reporting framework.

4 Changing Profile of the Accountant and Auditing Profession

With the evolution of new technologies, the accounting and auditing professions must adapt too. Accountants, who were traditionally viewed as record keepers and verification agents of financial information, will have a more strategic and management-oriented role (Smith, 2018). Nevertheless, the rapid digitization and greater technological integration throughout the profession pose systematic risks to the viability of the profession in the medium-term. Therefore, as the accounting profession evolves and becomes more technology-dependent in nature, the willingness to learn and the ability to leverage existing competencies to address new problems will be crucial (Smith, 2018). Similarly, Vasarhelyi et al. (2015) and Cao et al. (2015) highlight that Big Data will cause significant changes in accounting education, pointing out the increase in statistical and IT content in accounting and auditing curricula.

Thus, the technological discruption might have three major implications for the accounting and auditing profession: role diversification, emergence of new sectors, and changes in the accounting and auditing curriculum.

4.1 Role Diversification

As repetitive tasks might soon be almost all automated, accountants are expected to play more central roles within businesses. In addition, there will be an increased demand for accountants to read and understand the kind of data resulting from the incorporation of new technologies into business and accounting processes such as Big Data, IoT, and blockchain. Thus, accountants would be required to adquire more analytical but also managerial skills. In addition, recent research on the type of skills necessary to cover the demand of the current labour market affected by the evolution of new technologies reveal that there are three main areas of skills: ICT skills, business, and transversal skills (CHAISE, 2021a). Obviously, the level of knowledge in each skills area (ICT, business, transversal) depends on the functional role of the candidate and not all functional roles would require an expert level in ICT skills. Nevertheless, a general understanding of technology or beginner level ICT skills might be more commonly required.

4.2   Emergence of New Sectors

Something similar to the emergence of Fintech sector, which is a combination of Finance and Technology, might occur in the Accounting and Auditing field as well. There are already numerous projects related to AI and blockchain in Auditing which requires a collaboration between accounting/auditing experts and IT specialists. Thus, it is probably a matter of time until Audit-tech sector (or similar) will emerge. It has definitely started to take shape.

4.3   Changes in Accounting and Auditing Curriculum

The recent survey regarding blockchain also points out that there is a higher demand for blockchain-skilled candidates than the offer can satisfy (CHAISE, 2021a). This is due to a slow responsiveness of formal education on current labour market needs. Therefore, companies are often forced to provide in-house training, which is time-consuming and costly. Similarly, candidates opt for informal educational activities e.g. through learning platforms like Coursera, Udemy, or private academies. Thus, it is necessary that formal education adopt a more proactive approach and extend traditional accounting and auditing curriculum with courses related to blockchain, smart contracts, AI and machine learning, etc.

Conclusions

In this original monothematic work of scientific-exploratory character I combined the results from my recent empirical studies related to ADM disclosure and AI reporting framework with a comprehensive literature and technology review. Thus, the findings obtained from literature review, desk research, focus groups, questionnaires and statistical analysis were used to critically analyse, evaluate, and synthesize the knowledge related to new technologies such as AI, IoT, and blockchain. First, a level of disruption caused by these new technologies is discussed, analysing technology functionalities, implications and challenges for accounting and business. Consequently, the implications of these technologies were put into a context of CDR, where I also discussed the latest initiatives in AI ethics and trends in AI disclosure, which still has is still voluntary. Economic-based and socio-political theories were applied to help understand the motivation of companies for voluntary AI disclosure. As our previous research outlined, companies struggle with AI reporting, as on one hand there is an increase stakeholder pressure to be transparent about AI usage, but on the other hand, there are no standards or guidelines related to what and how to report. Therefore, our proposal of a set of elements for AI disclosure, which has been developed based on a multi-stakeholder approach, might be considered a first step towards standardization in this area. The fourth chapter of this book is related to an idea of continuous reporting, which has emerged as implications of digital disruption related to the availability of tools such as AI, IoT, XBRL, blockchain, and corporate digital responsibility. Thus, it seems to be a natural evolution of corporate reporting. Last but not least, the changing profile of accountant and auditor profession was discussed.

References

Allison, I. (2015). Deloitte, Libra, Accenture: The work of auditors in the age of Bitcoin 2.0 technology. International Business Times. http://www.ibtimes.co.uk/deloitte-libra-accenture-work-auditors-age-bitcoin-2-0-technology-15159 32 Accessed 1 April 2018.

Algorithmwatch (2020). AI Ethics Guidelines Global Inventory. https://inventory.algorithmwatch.org Accessed 2 March 2021.

Angwin, J., Larson, J., Mattu, S., & Kirchner, L. (2016). Machine bias. Pro Publica. https://www.propublica.org/article/machine-bias-risk-assessments-in-criminal-sentencing Accessed 2 April 2017.

Aizawa, A. (2003). An information-theoretic perspective of tf-idf measures. *Information Processing and Management*, 39 (1), 45–65. https://doi.org/10.1016/S0306-4573(02)00021-3

Ashworth, L., & Free, C. (2006). Marketing dataveillance and digital privacy: Using theories of justice to understand consumers' online privacy concerns. *Journal of Business Ethics, 67*(2), 107–123.

Beke, F. T., Eggers, F., & Verhoef, P. C. (2018). Consumer informational privacy: Current knowledge and research directions. *Foundations and Trends in Marketing, 11*(1), 1–71.

Bonsón, E., & Bednárová, M. (2018). Blockchain y los registros contables consensuados compartidos (RC3). *Revista AECA*, 123, 4-5.

Bonsón, E., & Bednárová, M. (2019). Blockchain and its implications for Accounting and Auditing. *Meditari Accountancy Research*, 27 (5), 725-740. https://doi.org/10.1108/MEDAR-11-2018-0406.

Bonsón, E., & Bednárová, M. (2021a). Hacia una inteligencia artificial socialmente responsable: principios éticos y regulación. *Revista AECA*, 131, 8-10.

Bonsón, E., & Bednárová, M. (2021b). Presentación de la información relacionada con la utilización de la Inteligencia Artificial en el Estado de Información no Financiera. *Opinión Emitida AECA*.

Bonsón, E., & Bednárová, M. (2022). Artificial Intelligence Disclosures in Sustainability Reports: Towards an Artificial Intelligence Reporting Framework. Chapter in the book: *Digital Transformation in Industry*. Springer. 10.1007/978-3-030-94617-3_27.

Bonsón, E., Bednárová, M., Perea, D., & Alejo, V. (2022a) ADM Disclosure. Working paper.

Bonsón, E., Bednárová, M., & Perea, D. (2022b). Blockchain Disclosure. Working paper.

Bonsón, E., & Ortega, M. (2019). Big Data, Inteligencia Artificial y Data Analytics (BIDA): un foro necesario promovido por AECA. *Revista AECA, 125,* 11-13.

Bonsón, E., Lavorato, D., Lamboglia, R., & Mancini, D. (2021a). Artificial Intelligence activities and ethical approaches in leading listed companies in the European Union. *International Journal of Accounting Information Systems, 43.*

Bonsón, E., Alejo, V., & Lavorato, D. (2021b). Artificial Intelligence Disclosure in the Annual Reports of Spanish IBEX-35 Companies (2018-2019). *Digital Transformation in Industry, Lecture Notes in Information Systems and Organisation,* (pp. 147-155). Springer. 10.1007/978-3-030-73261-5_14.

Bonsón, E., & Bednárová, M. (2015). CSR reporting practices of Eurozone companies. *Spanish Accounting Review,* 18(2), 182-193.

Benaich, N., & Hogarth, I. (2020) State of AI Report. https://docs.google.com/presentation/d/1ZUimafgXCBSLsgbacd6-a-dqO7yLyzIl1ZJbiCBUUT4/edit#slide=id.g557254d430_0_0 Accessed 2 March 2021.

Buterin, V. (2014). Ethereum White Paper: A next-generation smart contract and decentralized application platform. https://github.com/ethereum/wiki/wiki/White-Paper Accessed 3 May 2018.

Bynum, T. W. (2001). Computer ethics: Its birth and its future. *Ethics and Information Technology, 3*(2), 109–112.

CHAISE (2021a) Study on Blockchain labour market characteristics. https://chaise-blockchainskills.eu/wp-content/uploads/2021/05/D2.2.1-Study-on-Blockchain-labour-market-characteristics.pdf Accessed 20 April 2022.

CHAISE (2021b) Study on skills mismatches in the Blockchain sector. https://chaise-blockchainskills.eu/wp-content/uploads/2021/11/CHAISE_WP2_D2.5.1_Study-on-Skills-Mismatches-in-the-bloc kchain-sector.pdf Accessed 20 April 2022.

Cai, Y., & Zhu, D. (2016). Fraud detections for online businesses: A perspective from blockchain technology. *Financial Innovation*, 2(1), 20.

Campbell, D. Craven, B., & Shrives, P. (2003). Voluntary social reporting in three FTSE sectors: a comment on perception and legitimacy. *Accounting, Auditing & Accountability Journal,* 16(4), 558-581.

Cao, M., Chychyla, R., & Stewart, T. (2015). Big Data analytics in financial statement audits. *Accounting Horizons,* 29 (2).

Cath, S., (2018). Governing artificial intelligence: ethical, legal and technical opportunities and challenges. *Philosophical transactions of the royal society a-mathematical physical and engineering sciences.* https://doi.org/10.1098/rsta.2018.0080.

Chatterjee, S., Moody, G., Lowry, P. B., Chakraborty, S., & Hardin, A. (2015). Strategic relevance of organizational virtues enabled by information technology in organiza- tional innovation. *Journal of Management Information Systems,* 32(3), 158–196.

Connelly, B. L., Certo, S. T., Ireland, R. D., & Reutzel, C. R. (2011). Signaling theory: A review and assessment. *Journal of management*, 37(1), 39-67.

Culnan, M. J., & Bies, R. J. (2003). Consumer privacy: Balancing economic and justice considerations. *Journal of Social Issues,* 59(2), 323–342.

Conpolicy (2021). The Institute of Consumer Policy: Corporate Digital Responsibility. https://www.conpolicy.de/en/topics/corporate-digital-responsibility/ Accessed 1 March 2021.

Cybersecurity Act (2021). Candidate EUCC scheme. https://european-accreditation.org/cybersecurity-act-candidate-eucc-scheme/ Accessed 2 March 2022.

Corea, F. (2019). *An Introduction to Data. Everything you need to know about AI, Big Data and Data Science.* Springer. https://link.springer.com/content/pdf/bfm%3A978-3-030-04468-8%2F1.pdf

Costa, I. D. (2021). Protecting individuals in a big data age: the opacity of the algorithm and automated decision-making. *RED-Revista Electronica de Direito*, 24(1), 33-82.

Cobbe, J., & Singh, J. (2020). Reviewable Automated Decision-Making. *Computer law and security review*, 39. https://doi.org/10.1016/j.clsr.2020.105475.

Cyert, R.M., & March, J.G. (1963). *A behavioral theory of the firm*. Englewood Cliffs, NJ: Prentice-Hall.

Dai, J., & Vasarhelyi, M. A. (2017). Toward Blockchain-Based Accounting and Assurance. *Journal of Information Systems*, 31(3), 5-21.

Davenport, T.H., & Harris, J.G. (2005). Automated decision making comes of age. *MIT Sloan Management Review,* 46(4), 83.

Deegan, C. (2002). Introduction: the legitimising effect of social and environmental disclosures a theoretical foundation. *Accounting, Auditing & Accountability Journal,* 15(3), 282-311.

Drew, J. (2017). Real talk about artificial intelligence and blockchain. *Journal of Accountancy*, 224(1), 1-6.

Deloitte (2019). Redesigning Corporate Responsibility. How Digitalization changes the role companies need to play for positive impacts on society. https://www2.deloitte.com/de/de/pages/innovation/contents/redesigning-corporate-responsibility.html Accessed 1 May 2021.

Dye, R. A. (2001). An evaluation of "essays on disclosure" and the disclosure literature in accounting. *Journal of accounting and economics*, *32(*1-3), 181-235.

European Commission (2018). Declaration of Cooperation on AI. https://ec.euro-pa.eu/jrc/communities/sites/jrccties/files/2018aideclarationatdigitaldaydocxpdf.pdf Accessed 5 January 2020.

European Commission (2019). Ethics Guidelines for Trustworthy AI. https://www.aepd.es/sites/default/files/2019-12/ai-ethics-guidelines.pdf Accessed 5 January 2020.

European Commission (2020). White Paper – On Artificial Intelligence – A European approach to excellence and trust. https://ec.europa.eu/info/sites/default/

files/commission-white-paper-artificial-intelligence-feb2020_en.pdf Accessed 1 April 2022.

European Commission (2021a). Proposal for a regulation of the European Parliament and of the Council. Laying down harmonized rules on Artificial Intelligence (Artificial Intelligence Act) and amending certain Union legislative acts. https://eur-lex.europa.eu/legal-content/EN/TXT/HTML/?uri=CELEX: 52021 PC0206&from=EN Accessed 1 April 2022.

European Commission (2021b) Communication from the Commission to the European Parliament, the European Council, the European Economic and Social Committee and the Committee of the Regions. Fostering a European approach to AI.

European Commission (2022). Overview of EU funded blockchain related projects. https://digital-strategy.ec.europa.eu/en/news/overview-eu-funded-blockchain-related-projects Accessed 15 April 2022.

European Parliament (2014). Directive 2014/95/EU of the European Parliament and of the Council of 22 October 2014 amending Directive 2013/34/EU as regards disclosure of non-financial and diversity information by certain large undertakings and groups.

EFRAG (2021). European Reporting Lab @ EFRAG. Proposals for a relevant and dynamic EU sustainability reporting standard-setting. https://www.efrag.org/Assets/Download?assetUrl=%2Fsites%2Fwebpublishing%2FSiteAssets%2FEFRAG%2520PTF-NFRS_MAIN_REPORT.pdf Accessed 1 April 2022.

Eur-lex (2013) Transparency Directive. https://eur-lex.europa.eu/legal-content/EN/NIM/?uri=celex:32013L0050) Accessed 5 March 2020.

ESMA (2019). ESEF. Commission Delegated Regulation (EU) 2019/815, ESMA XBRL. https://www.esma.europa.eu/esef-example-annual-financial-report Accessed 1 April 2022.

Fjeld, J., Achten, N., Hilligoss, H., Nagy, A., & Srikumar, M. (2020). *Principled artificial intelligence: mapping consensus in ethical and rights-based approaches to principles for AI.* Berkman Klein Center Research Publication. https://doi.org/10.2139/ ssrn.3518482.

Floridi, L. (2010). Ethics after the information revolution. In L. Floridi (Ed.). *Cambridge Handbook of Information and Computer Ethics* (pp. 3–19). Cambridge: Cambridge University Press.

Floridi, L. (2019). Translating principles into practices of digital ethics: five risks of being unethical. *Philosophy and Technology*, 32, 185-193. https://doi.org/10.1007/s13347-019-00354-x.

Franks, B. (2012). *Taming the Big Data Tidal Wave: Finding Opportunities in Huge Data Streams with Advanced Analytics.* New York, NY: Wiley.

Future of Life Institute (2017). Beneficial AI. https://futureoflife.org/bai-2017/ Accessed 1 April 2022.

Gandía, J.L. (2018). Tecnología, contabilidad y blockchain, retos y oportunidades para el siglo XXI. In *Retos de la contabilidad y la auditoría en la economía actual* (pp. 341-351). Valencia: Universidad de Valéncia.

Gartner (2019). Gartner Survey Shows 37 Percent of Organizations Have Implemented AI in Some Form. https://www.gartner.com/en/newsroom/press-releases/2019-01-21-gartner-survey-shows-37-percent-of-organizations-have Accessed 1 July 2021.

Gerrity, T.P. (1970). *The design of man-machine decision systems* (Ph.D. Dissertation), Sloan School of Management, Massachusetts Institute of Technology. http://hdl.handle.net/1721.1/92990.

Gerwien, M., Jungmann, A., & Vosswinkel, R. (2021). Towards Situation-Aware Decision-Making for Automated Driving. *7th International Conference on Automation, Robotics and Applications* (ICARA), pp. 185-189.

Google (2018). Artificial Intelligence at Google. https://ai.google/principles/ Accessed 20 March 2022.

Grigg, I. (2005). Triple Entry Accounting. http://iang.org/papers/triple_entry.html Accessed 15 March 2021.

Grishina, T.G. (2012). Likelihood Substantiation and decision-making at management of the automated manufacture. *Mekhatronika, avtomatizatsiya, upravlenie,* 1, 48-52.

Greene, D., Hoffmann, A., & Stark, L. (2019). A Critical Assessment of the Movement for Ethical Artificial Intelligence and Machine Learning. *Pro-*

ceedings of the 52nd Hawaii International Conference on System Sciences. https://hdl.handle.net/10125/59651ISBN: 978-0-9981331-2-6(CC BY-NC-ND 4.0)Page 2122 Accessed 12 March 2021

Guarda, D. (2018). The future of finance – Fintech, AI, Blockchain. https://www.slideshare.net/dinisguarda/the-future-of-finance-fintech-ai-blockchain-by-dinis-guarda Accessed 12 April 2021.

Gorry, A., & Scott Morton, M.S. (1971). A framework for information systems. *Sloan Management Review,* 13(1), 56–79.

Hagendorff, T. (2020). The ethics of AI ethics: an evaluation of guidelines. *Minds and Machines,* 30, 99-120. https://doi.org/10.1007/ s11023-020-09517-8

Hedgethink (2016). Blockchain and IoT centralised and decentralised tech. https://www.hedgethink.com/fintech-revolution-ai-iot-driven-blockchain-tsunami/screen-shot-2016-11-16-at-10-03-44/ Accessed 15 May 2021.

Hickok, M. (2020). Lessons learned from AI ethics principles for future actions. *AI and Ethics.* Springer Nature Switzerland AG 2020. https://doi.org/10.1007/s43681-020-00008-1

Holsti, O.R. (1969), *Content Analysis for the Social Sciences and Humanities.* Addison-Wesley, Reading, MA.

IBM (2018). IBM's multidisciplinary, multidimensional approach to trustworthy AI. https://www.ibm.com/artificial-intelligence/ethics Accessed 20 March 2022.

IEEE Global Initiative on Ethics of Autonomous and Intelligent Systems (2016). https://standards.ieee.org/wp-content/uploads/import/documents/other/ead1e.pdf Accessed 5 May 2021.

Isensee, C., Teuteberg, F., Griese, K.M, & Topi, C. (2020). The relationship between organizational culture, sustainability, and digitalization in SMEs: A systematic review. *Journal of Cleaner Production,* 275, 122944. 10.1016/j.jclepro.2020.122944.

Jobin, A., Ienca, M., & Vayena, E. (2019). The global landscape of AI ethical guidelines. *Nat. Mach. Intell.* 1, 389-399.

Jayachandran, P. (2017). The difference between public and private blockchain. https://www.ibm.com/blogs/blockchain/2017/05/the-difference-between-public-and-private-blockchain/ Accessed 25 April 2021.

Kehr, F., Kowatsch, T., Wentzel, D., & Fleisch, E. (2015). Blissfully ignorant: The effects of general privacy concerns, general institutional trust, and affect in the privacy calculus. *Information Systems Journal, 25*(6), 607–635.

Kokina, J., Mancha, R. y Pachamanova, D. (2017). Blockchain: Emergent Industry Adoption and Implications for Accounting. *Journal of Emerging Technologies in Accounting,* 14 (2), 91-100.

Kolbe, R.H., & Burnett, M.S. (1991). Content analysis research: an examination of application with directives for improving research reliability and objectivity. *Journal of Consumer Research,* 18 (2), 243-250.

Landauer, T.K., Foltz, P.W., & Laham, D. (1998). An introduction to latent semantic analysis. *Discourse Processes,* 25 (2–3).

Lobschat, L., Mueller, B., Eggers, F., Brandimarte, L., Diefenbach, S., Kroschke, M., & Wirtz, J. (2021). Corporate Digital Responsibility. *Journal of Business Research*, 122, 875-888.

Lu, J., & Wang, J. (2021). Corporate governance, law, culture, environmental performance and CSR disclosure: A global perspective. *Journal of International Financial Markets, Institutions and Money*, 70, 101264.

Liyanaarachchi, G., Deshpande, S., & Weaven, S. (2020). Market-oriented corporate digital responsibility to manage data vulnerability in online banking. *International Journal of Bank Marketing,* 10.1108/IJBM-06-2020-0313.

Mainelli, M., & Smith, M. (2015). Sharing ledgers for sharing economies: An exploration of mutual distributed ledgers (aka blockchain technology), The Journal of Financial Perspectives, 3 (3), 38–69.

Marwala, T., & Xing, B. (2018). Blockchain and Artificial Intelligence. University of Johannesburg, Auckland Park.

Mason, R. O. (1986). Four ethical issues of the information age. *MIS Quarterly,* 10(1), 5–12.

McAfee, A., & Brynjolfsson, E. (2012). Big data: The management revolution. *Harvard Business Review*, 60–66. https://hbr.org/2012/10/big-data-the-mana gement-revolution Accessed 20 March 2022.

McKinsey Global Institute (2011). Big data: The next frontier for innovation, competition, and productivity. https://www.mckinsey.com/business-func tions/mckinsey-digital/our-insights/big-data-the-next-frontier-for-innovation Accessed 10 April 2021.

Microsoft (2018). Responsible AI. https://www.microsoft.com/en-us/ai/responsi ble-ai?activetab=pivot1%3aprimaryr6 Accessed 20 March 2022.

Moffit, K., & Vasarhelyi, M.A. (2013). AIS in an Age of Big Data. *Journal of Information Systems.* 27 (2), 1-19.

Moor, J. H. (1985). What IS computer ethics? *Metaphilosophy, 16*(4), 266–275.

Mougiakou, E., Papadimitriou, S., & Virvou, M. (2019). Automated Decision Making and Personal Data Protection in Intelligent Tutoring Systems: Design Guidelines. *Knowledge-based software engineering,* 108, 231-241.

Minichiello, N. (2015). Deloitte launches Rubix, a one stop blockchain software platform. https://bravenewcoin.com/news/deloitte-launches-rubix-a-one-stop-blockchain-software-platform/ Accessed 15 April 2021.

Nakamoto, S. (2008). Bitcoin: A Peer-to-Peer Electronic Cash System. https://bit coin.org/bitcoin.pdf Accessed 10 March 2021.

Nofer, M, Gomber, P., Hinz, O., & Schiereck, D. (2017). Blockchain. *Business and Information Systems Engineering,* 59 (3), 183-187.

O'brien, H.M. (2016). The internet of things. *Journal of Internet Law*, 19 (12), 1-20.

Oracle (2018). Transformational Technologies Today. How IoT, AI, and block-chain will revolutionize business. http://www.oracle.com/us/solutions/cloud/ tt-technologies-white-paper-4498079.pdf Accessed 1 March 2020.

Ølnes, S., Ubacht, J., & Janssen, M. (2017). Blockchain in government: Benefits and implications of distributed ledger technology for information sharing. *Government Information Quarterly,* 34(3), 355-364.

Osburg, T. (2017). Sustainability in a Digital World Needs Trust. In: *Sustainabi-lity in a Digital World* (pp. 3-19). Springer International Publishing.

Palfreyman, J. (2015). Blockchain for government? https://medium.com/@JohnP 261/blockchain-for-government-184ace54756d Accessed 2 December 2018.

Power, D.J., Heavin, C., & Keenan, P. (2019). Decision systems redux. *Journal of Decision Systems,* 28(1), 1-18.

PWC, 2019. Corporate Digital Responsibility and Digital Ethics. https://www.pwc.de/en/sustainability/corporate-digital-responsibility-and-digital-ethics.html Accessed 1 April 2021.

R Core Team, 2018. R: A Language and Environment for Statistical Computing. https://www.r-project.org/.

Robbins, S. (2020). AI and the path to envelopment: knowledge as a first step towards the responsible regulation and use of AI-powered machines. *AI & Society,* 35, 391–400.

Rozario, A. & Vasarhelyi, M.A. (2018). Auditing with Smart Contracts. *The International Journal of Digital Accounting Research*, 18, 1-27. 10.4192/1577-8517-v18_1.

Sangdon, P. (2017). A constitutional overview of right to explanation of automated decision-making algorithm. *Constitutional Law*, 23(3), 185-218.

Sedia (2021). Carta de derechos digitales. XXIII Derechos ante la Inteligencia Artificial. https://portal.mineco.gob.es/RecursosArticulo/mineco/ministerio/participacion_publica/audiencia/ficheros/SEDIACartaDerechosDigitales.pdf Accessed 1 February 2022.

Seufert, M., Lange, S., & Meixner, M. (2016). Automated Decision Making for the Multi-objective Optimization Task of Cloud Service Placement. *28th International teletraffic congress (ITC 28),* 2, 16-21.

Schwab, K. (2016). The Fourth Industrial Revolution: what it means, how to respond, World Economic Forum. https://www.weforum. org/agenda/2016/01/the-fourth-industrial-revolution-what-it-means-and-how-to-respond/ Accessed 1 March 2018.

Simon, H. (1960). *The new science of management decision.* New York: Harper & Brothers Publishers. https://babel.hathitrust.org/cgi/ pt?id=uc1.b3376401;view =1up;seq=10.

Sgantzos, K., & Grigg, I. (2019). Artificial Intelligence Implementations on the Blockchain. Use Cases and future applications. *Future Internet,* 11(8), 170. https://doi.org/ 10.3390/fi11080170.

Smith, S. S. (2018). Digitization and Financial Reporting – How Technology Innovation May Drive the Shift toward Continuous Accounting. *Accounting and Finance Research,* 7 (3), 240-250.

Smith, A.N., Fisher, E., & Yongjian, C. (2012). How does brand-related user-generated content differ across YouTube, Facebook, and Twitter? *Journal of Interactive Marketing,* 26(2), 102-113.

Stein, S.S. (2018). Digitization and Financial Reporting – How Technology Innovation May Drive the Shift toward Continuous Accounting. *Accounting and Finance Research*, 7(3), 240-250.

Suchman, M.C. (1995). Managing legitimacy: Strategic and institutional approaches. *Academy of Management Review*, 20(3), 571-610.

Swan, M. (2015). *Blockchain: Blueprint for a New Economy.* O'Reilly Media, Inc.

Sutton, S.G., Holt, M., & Arnold, V. (2016). The reports of my death are greatly exaggerated-Artificial intelligence research in accounting. *International Journal of Accounting Information Systems*, 22, 60-73.

Tapscott, D., & Tapscott, A. (2016). The impact of blockchain goes beyond financial services. *Harvard Business Review.* https://hbr.org /2016/05/the-impact-of-the-blockchaingoes-beyond-financial-servic es Accessed 1 February 2019.

Telefonica (2018). AI Principles of Telefonica. https://www.telefonica.com/wp-content/uploads/sites/7/2021/11/principios-ai-eng-2018.pdf Accessed 20 March 2022.

Tenomad (2018). Block structure. https://tenomad.com/a-blockchain-in-200-lines-of-code/ Accessed 25 April 2022.

Thiprungsri, S., & Vasarhelyi, M. (2011). Cluster analysis for anomaly detection in accounting data: An audit approach. *International Journal of Digital Accounting Research*, 11, 69-84. 10.4192/1577-8517-v11_4.

Thorun, C. (2018). Corporate Digital Responsibility – Unternehmerische Verantwortung in der digitalen Welt. In: Fallstudien zur Digitalen Transformation (pp. 174-191). SpringerGabler.

Tolan , S., Miron, M., Gómez, E., & Castillo, C. (2019). Why Machine Learning May Lead to Unfairness: Evidence from Risk Assessment for Juvenile Justice in Catalonia. Conference paper: International Association for AI and Law. Montreal, Canada. DOI:10.1145/3322640.3326705.

Vakkuri, V., Kemell, K.K., Kultanen, J., & Abrrahamsson, P. (2020). The current state of Industrial Practice in Artificial Intelligence Ethics. *IEEE Software,* 37(4), 50-57. 10.1109/MS.2020.2985621.

Vasarhelyi M. A., Alles, M. G., & Williams, K. T. (2010). Continuous Assurance for the Now Economy. Sydney, Australia: Institute of Chartered Accountants in Australia.

Vasarhelyi, M.A., Kogan, A., & Tuttle, B.M. (2015). Big Data in Accounting: An Overview. *Accounting Horizons,* 29(2), 381-396.

Vidgen, R., Hindle, G., & Randolph, I. (2020). Exploring the ethical implications of business analytics with a business ethics canvas. *European Journal of Operational Research,* 281 (3), 491-501

Wanden-Berghe, J.S., Bednárová, M., & Fernández, E. (2019). *La tecnología blockchain y sus implicaciones en el ámbito empresarial. Documentos AECA: Nuevas tecnologías y contabilidad.* Documento N. 15. Asociación española de contabilidad y administración de empresas (AECA), Madrid (Spain).

Wanden-Berghe, J.S., Bednárová, M., & Fernández, E. (2020). La disrupción tecnológica y su impacto en el ámbito económico y de la información. *Revista AECA,* 129, 21-25.

XBRL (2021). ESEF implementation. https://www.xbrl.org/news/esef-implementation-begins-in-europe-and-esap-promises-easy-data-access/ Accessed 20 March 2022.

Zhou, S., Simnett, R., & Green, W. (2017). Does integrated reporting matter to the capital market? *Abacus,* 53(1), 94-132.